Topics of Thought

Topics of Thought
The Logic of Knowledge, Belief, Imagination

Francesco Berto

OXFORD
UNIVERSITY PRESS

OXFORD
UNIVERSITY PRESS

Great Clarendon Street, Oxford, OX2 6DP,
United Kingdom

Oxford University Press is a department of the University of Oxford.
It furthers the University's objective of excellence in research, scholarship,
and education by publishing worldwide. Oxford is a registered trade mark of
Oxford University Press in the UK and in certain other countries

© Francesco Berto 2022

The moral rights of the author have been asserted

First Edition published in 2022
Impression: 1

Published in the United States of America by Oxford University Press
198 Madison Avenue, New York, NY 10016, United States of America

British Library Cataloguing in Publication Data
Data available

Library of Congress Control Number: 2022934444

ISBN 978-0-19-285749-1

DOI: 10.1093/oso/9780192857491.001.0001

Printed and bound in the UK by
TJ Books Limited

Links to third party websites are provided by Oxford in good faith and
for information only. Oxford disclaims any responsibility for the materials
contained in any third party website referenced in this work.

Contents

To the memory of my dad

Acknowledgments

This book is one main outcome of the 2017-2022 research project *The Logic of Conceivability: Modelling Rational Imagination with Non-Normal Modal Logics*, funded by the European Research Council (ERC) under the Horizon 2020 Research and Innovation Programme, Grant Agreement No. 681404, and co-hosted by the Department of Philosophy at the University of St Andrews and by the Institute for Logic, Language and Computation (ILLC) at the University of Amsterdam. I am grateful to my colleagues at both institutions for surrounding me with two perfect academic environments during my research.
At the beginning of 2014, I moved from Scotland to Amsterdam: I arrived at the ILLC with little knowledge of epistemic logic and no interest in epistemology. By the end of 2018, when I returned to my beloved Scotland to join St Andrews, my exposure to the ILLC's great tradition in epistemic logic had caused my interests to expand in ways this book makes apparent.

My biggest debt is to the Logic of Conceivability Gang, for helping me to shape most of the ideas the book plays with: Peter Hawke and Aybüke Özgün (who have also co-authored its Chapters 2 and 4 and, respectively, 7 and 8), Karolina Krzyżanowska, Tom Schoonen, Anthia Solaki, Chris Badura, and Thomas Ferguson.

In 2020 I presented most of the book's contents twice, in two series of seminars delivered online (you know why) to heroic audiences, during my honorary Chaire Mercier Lectures 'at' the Université Catholique de Louvain, and 'at' the 9th Summer School in Philosophy of the University of

Hamburg. Between 2016 and 2021, I presented individual Chapters, or the papers they are based on, in a number of venues: the seminar of the Arché Logic Group at the University of St Andrews; the Non-Categorical Thinking Workshop at the Center for Logic Language and Cognition, University of Turin; the Kyoto Workshop on Paraconsistency and Dialetheism at the University of Kyoto; the Tokyo Forum for Analytic Philosophy at the University of Tokyo; the workshop on Imagination and Modality at the University of Padua; the conference on Conceivability and Modality at the University of Rome-La Sapienza; the 9th European Conference of Analytic Philosophy at the University of Munich; the workshop on Doxastic Agency and Epistemic Logic at the Ruhr University of Bochum; the conference on the Philosophy of Imagination, again at the University of Bochum; the Conference on Reasoning in Social Contexts at the Royal Netherlands Academy of Arts and Sciences; the Philosophy of Language and Logic Seminar at the University of Milan; a seminar at the Dalle Molle Institute for Artificial Intelligence in Lugano; the Logica 2018 conference in the Czech Republic; the conference on Models of Bounded Reasoning in Individuals and Groups at the Lorentz Center in Leiden; the 41st International Wittgenstein Symposium in Kirchberg am Wechsel; a conference at the Royal Society of Edinburgh; the Philosophy Seminar at the University of Stirling; a seminar for the MUMBLE Research Group at the University of Turin; the Philosophical Reflectorium at the University of St Andrews; a seminar at the Institute of Philosophy, University of London.

As you may have guessed from such a long list, way too many people have come up with good questions, suggestions, and criticisms, for me to be able to thank everyone individually. I'll give it a try, with apologies to those I have forgotten: Arif Ahmed, Maria Aloni, Holger Andreas, Sergei Artemov, Magdalena Balcerak-Jackson, Alexandru Baltag, Johan van Benthem, Corine Besson, Sarah Broadie, Jessica Brown, Claudio Calosi, Ilaria Canavotto, Max Carrara, Joan Casas Roma, Roberto Ciuni, Damiano Costa, Aaron Cotnoir, Enzo Crupi, Marcello D'Agostino, John Divers, Mike Dunn, Marie Duží, Paul Egré, Peter van Emde Boas,

Jorge Ferreira, Camille Fouché, Lello Frascolla, Ale Giordani, Martin Glazier, Valeria Giardino, Matt Green, Patrick Greenough, Davide Grossi, Katherine Hawley, Levin Hornischer, Hykel Hosni, Andrea Iacona, Luca Incurvati, Manuel Gustavo Isaac, Bruno Jacinto, Mark Jago, Bjørn Jespersen, Amir Kiani, Amy Kind, Stephan Krämer, Michiel van Lambalgen, Dan Lassiter, Miguel Leon Untiveros, Hannes Leitgeb, Martin Lipman, Jiqi Liu, Tito Magri, Diego Marconi, Neri Marsili, Peter Milne, Sanjay Modgil, Sarah Moss, Bence Nanay, Antonio Negro, Daniel Nolan, Sebastian Obrist, Sergei Odintsov, Hitoshi Omori, Naomi Osorio-Kupferblum, David Over, Walter Pedriali, Matteo Plebani, Graham Priest, Simon Prosser, Thomas Randriamahazaka, Stephen Read, François Recanati, Greg Restall, Floris Roelofsen, Robert van Rooij, Stefan Roski, Gill Russell, Ben Sachs, Pierre Saint-Germier, Paolo Savino, Kevin Scharp, Ben Schnieder, Katrin Schulz, Sonja Smets, Justin Snedegar, Roy Sorensen, Beppe Spolaore, Margot Strohminger, Michael Stuart, Neil Tennant, Peter Verdee, Alberto Voltolini, Heinrich Wansing, Zach Weber, Tim Williamson, Crispin Wright, Steve Yablo. I am also grateful to the staff at Oxford University Press; to Peter Momtchiloff, for being such a wise editor and for providing terrific anonymous reviewers (thanks to you, too!) for the book manuscript; and to Mark Jago (again) for allowing me to format this book using an elegant LaTeX template of his design.

Various parts of the book rely on previous work. Chapter 1 is new. Chapter 2 is mostly new, but also uses material from an unpublished manuscript co-authored with Peter Hawke and Levin Hornischer. Chapter 3 is mostly new, but also includes (a recap of) ideas that showed up in papers on which the following Chapters 4 to 6 are based. These go as follows: Chapter 4 is based on 'Knowability Relative to Information', *Mind* 2021 (with Peter Hawke). Chapter 5 is based on 'Equivalence in Imagination', in *Epistemic Uses of Imagination* (Routledge 2021, edited by Amy Kind and Chris Badura), though it also includes ideas taken from the older 'Aboutness in Imagination', *Philosophical Studies* 2017. Chapter 6 includes ideas from 'Simple Hyperintensional Belief Revision', *Erkenntnis* 2018, as well as new material, and fixes some claims from that 2018 paper which, I now think, were not quite right. Chapter 7 is based on an unpublished manuscript, provisionally titled 'The Logic of Framing Effects' and co-authored with Aybüke Özgün. Chapter 8 is based on 'Indicative Conditionals: Probabilities and Relevance', *Philosophical Studies* 2021, co-authored with Aybüke again. I am grateful to all the editors and publishers for permission to re-use material from the published works.

Finally: thanks mum, dad, Val, and Mabel, for keeping me from thinking too much about *thinking about*.

1

What Thoughts Are About

We think about this and that: when one thinks that Mary is happy, one's thought is about how Mary is doing. When one supposes that the markets will fall, one's supposition is about what will happen with the markets. When one believes that John is tall and handsome, one's belief is about John's height and looks. What is the logic of such thoughts? This book begins to explore the idea that, to answer the question, one has to take that 'about' at face value.

Aboutness is 'the relation that meaningful items bear to whatever it is that they are *on* or *of* or that they *address* or *concern*' (Yablo 2014, 1). This is their *subject matter* or, as I will also say, their *topic*. Research on aboutness and subject matter has been burgeoning in the last decades, thanks to the works of philosophers and logicians like David Lewis (1988a,b), Ken Gemes (1994, 1997), Lloyd Humberstone (2008), Stephen Yablo, Kit Fine (2016a, 2017), Peter Hawke (2018), Friederike Moltmann (2018), Arthur Schipper (2018, 2020), and more.

Such works address aboutness mostly, though not only, as a feature of those meaningful items which are pieces of language – in particular, declarative sentences. However, mental states can be meaningful, too. One traditional name for such meaningfulness is 'intentionality': the feature that some of our mental states have, of being directed towards objects, situations, or states of affairs. The book deals with

Topics of Thought: The Logic of Knowledge, Belief, Imagination.
Francesco Berto, Oxford University Press. © Francesco Berto 2022.
DOI: 10.1093/oso/9780192857491.003.0001

propositional or *de dicto* intentional states: states of the mind which are generally, though not universally, understood by philosophers as having propositions as their contents, and as being directed towards the situations the propositions are concerned with. Thus, such mental states are often called 'propositional attitudes': they are recorded linguistically by verbs taking sentential complements, such as 'believes (that)', 'knows (that)', 'imagines (that)', 'supposes (that)', 'is informed (that)', used to ascribe attitudes towards propositions, or towards what makes the propositions true. I will often use the term 'thoughts' as a cover-all for such states.[1]

The book explores a new approach to the logic of thought – a new, unified way of answering the question: given that one thinks (believes, knows, etc.) that φ, what other ψs does one think (believe, know, etc.) by the logic of one's thought? Under which logical operations is one's thought *closed*?

While addressing the question, the book also tries to show that such an approach has a lot to give to philosophy, in areas that range from mainstream epistemology (dogmatism, scepticism, fallibilism), to suppositional thinking and the cognitive role of imagination, to belief management and revision, probabilistic thinking, and conditionality.

---

[1] It would have been nice if I'd had something to say on the connections between aboutness, topicality, and *de re* intentional mental states – states directed towards objects, and recorded linguistically by verbs taking noun-phrase complements: 'Paul loves Peter', 'Mary admires Frodo', 'Carlos imagines the winged horse', 'The Greeks worshipped Zeus'. However, the connections between topicality and *de dicto* intentionality gave me plenty of work already for an initial exploration. Besides, *de re* intentionality may be especially related to the topicality of the subsentential components of sentences, and I think it is fair to say that this is currently a less developed area of aboutness research, though see, e.g., Hawke (2018), Badura (2021b), for promising ideas. Notorious puzzles of intentionality in the *de re* ballpark involve putative failures of the Substitutivity of Identity (or, more accurately, Substitutivity of (Rigid) Co-Referential Terms: Dave fears Jack the Ripper, but Dave doesn't fear his neighbour John; unbeknownst to him, John is Jack the Ripper); and problems concerning existence and existential generalization (Mary admires Frodo; does there exist, therefore, someone whom Mary admires?). I (still) think that the best way to address the latter kind of issue is to go Meinongian and admit nonexistent objects: see Zalta (1988); Berto (2012); Crane (2013); Priest (2016).

1.1   Closure

The most well-known logical answer to the question of closure comes from standard epistemic logic in the tradition of Hintikka (1962): we treat notions like *knows, believes, is informed that,* using normal modal logic. We represent attitude ascriptions via operators interpreted as quantifiers over possible worlds (ways things as a whole could be or have been), restricted from the viewpoint of a given world by a binary accessibility relation, R. We read R in different ways, depending on the operator: evidential indiscernibility, consistency with one's beliefs, etc. '$X\varphi$' ('The agent Xs that φ', where 'X' can ascribe knowledge, belief, etc.) is true at world w just in case φ is true at a bunch of worlds accessible via the relation R from w. By imposing simple conditions on R, we can then validate various principles characteristic of different modal systems, and which supposedly capture features of the relevant attitudes. Some conditions on R are more contentious than others. We generally agree that R should be reflexive for 'X' to be read as 'knows', given that knowledge is factive (one can only know true things); it shouldn't, for it to be read as 'believes' (one can have false beliefs). But we debate on whether R should be transitive, for we disagree on whether Positive Introspection should hold for knowledge or belief: does Xing that φ entail that one Xs that one Xs that φ?

Whatever constraints R may satisfy, this setting delivers unrestricted closure, i.e., full closure under logical consequence or entailment, for X *qua* normal modality: one Xs whatever is entailed by what one Xs. That's because, if one Xs that φ, then φ must be true at all possible worlds accessible via R. But that φ entails ψ is understood as meaning or implying that any world making φ true (in any model of our logic) will also make ψ true. And so one will also X that ψ. In particular, one always has the same attitude towards 'intensional', logical or necessary equivalents: if φ and ψ are true at the same possible worlds (of all models of our logic), one Xs that φ iff one Xs that ψ.

This represents highly idealized, logically omniscient agents (Fagin et al. 1995, ch. 9) that, for instance, can never retain inconsistent beliefs without trivially believing everything. The same happens in the mainstream AGM approach to belief revision (Alchourrón et al. 1985), as well as in epistemic logics that recapture AGM in a modal setting (Van Ditmarsch et al. 2008). Additionally, the standard Hintikkan framework is typically monotonic, in that it doesn't straightforwardly model, e.g., the idea that more information may result in less knowledge (Harman 1973; Brown 2018).

A shared feature of the operators introduced in this book, instead, is that they fail full closure: one sometimes Xs that φ without Xing a logically entailed or necessarily equivalent ψ. This is tied to what the propositions that φ and that ψ are (and, are not) about, and therefore one's thought is (and, is not) about. The aboutness of a *de dicto* intentional state, Xing that φ, should be suitably related to that of the proposition, P, which makes for the content of φ. And this book markets the idea that P may be understood, not just as the set of possible worlds where it is true, but also in terms of what it is about: its subject matter, or topic.[2] Therefore, the logic of intentionality must be topic-sensitive: topics explicitly feature in the semantics of the operators that represent attitude ascriptions. Due to such topic-sensitivity, the operators are hyperintensional (Berto and Nolan 2021): substitution of logical or necessary equivalents in their scope

---

[2]I said above that the contents of attitudes like belief or knowledge are not universally understood by philosophers to be propositions. Sarah Moss (2018) has recently advanced the original proposal that they be taken, instead, as sets of probability spaces, called probabilistic contents (roughly: a probability space is a set of possible worlds with an algebra of propositions, taken as sets of worlds in their turn, and a probability measure on the elements of the algebra). When it's about attitudes such as full (as opposed to graded) beliefs – the kind of attitudes we will focus on for the most part in our book – there is a correspondence between propositions and certain simple probabilistic contents: one can associate to each proposition P the set of probability spaces such that P is true at each world in their domain. Probabilities are idle for such probabilistic contents, for what they really represent is only a distinction between possible worlds (Moss 2018, 14). But even in this case, for Moss a full belief ascription doesn't ascribe belief in that simple content: 'John (fully) believes that Mary is happy' is used to convey that John has a probabilistic belief that's relevantly similar to the belief that it's certain that Mary is happy (Moss 2019). It would be an interesting task, although it's not one pursued in this book, to investigate how Moss' view may connect to topic-sensitive conceptions of propositional content.

can sometimes fail to preserve truth.

On the other hand, the operators introduced in the book are not logically anarchic. To use terminology going back, I think, to Dretske (1970), they are *semi-penetrating*: they penetrate through, or are closed under, some entailments or logical consequences, although they don't fully penetrate. One who advertises such operators, therefore, may end up in the uncomfortable position of facing both charges of excessive and of insufficient closure. For instance, defenders of full epistemic closure, i.e., full closure for the knowledge attitude, will stress that correct, competently carried out logical deduction is the safest way to preserve knowledge. One cannot be charged of committing the fallacy of logical deduction, Kripke (2011a) protests.

Vice versa, enemies of logical idealization will find counterexamples to any residual closure feature by looking at ordinary agents. If the point of avoiding logical omniscience is to represent realistic reasoners, why stop short of complete logical anarchy? Can't Joe Bloggs even fail to believe that φ while he believes that $\varphi \wedge \psi$ because he is, on occasion, cognitively incapacitated, otherwise busy, or so? Insofar as deduction happens in time, one could always get distracted and fail while in the business of applying Conjunction Elimination.[3]

Defending the specific validities and invalidities involving various topic-sensitive operators is one task carried out throughout this book. Here follow, instead, some general considerations concerning idealization.

---

[3]See Williamson (2000), 282 (he speaks of sudden death; I went for the less drastic distraction). Harman and Sherman (2004) criticize Williamson for assuming that 'deduction is a kind of inference, something one does', and claim that this 'confuses questions of implication with questions of inference' (495). As remarked by Holliday (2012), we use 'deduction', not only to refer to (what one might take as) abstract, perhaps structured, objects outside of spacetime, but also to refer to a human activity (as in 'Sherlock Holmes had made the right deduction one more time'): the mental process of drawing conclusions from premises in a certain way; and I think it's this latter sense that is relevant for Williamson's remark.

1.2   Logical Omniscience

Debates on logical omniscience in epistemic logic are at times phrased in terms of a rigid normative/descriptive dichotomy: is one dealing with ideal agents representing a normative standard, or is one describing ordinary thinkers? If the former, why not full closure? If the latter, why any closure at all? I think this way of putting things is misleading in a number of ways. Here's one: it confuses idealization with normativity. That logically omniscient, idealized agents are on top of certain logical consequences of their thoughts doesn't mean that *we* should be. φ entails $\varphi \vee \psi$, $\varphi \vee \psi \vee \chi$, and so on, and so logically omniscient agents who believe the first item in the series, believe the whole series. But given that one of *us* believes that it is raining, one is hardly rationally committed to coming to believe that either it is raining or there's life on Kepler-442b; that either it is raining or there's life on Kepler-442b or pigs can fly; and so on. Limited as we are in our time, attention, memory, and computational powers, we shouldn't waste our resources by engaging in such a stupid sequence of deductive moves. As argued, e.g., by Cherniak (1986), Harman (1986), rational thinkers must not 'clutter their mind' with pointless inferences.

Here's another confusion we need to avoid: we should not forget that we fail to be logically omniscient for very diverse and often orthogonal reasons. Sometimes we fail to have the right attitude towards some necessary truths simply because we lack some relevant empirical information: one is not informed that one's neighbour John is Jack the Ripper although one has no doubts on John's self-identity. We sometimes know some necessary truths because they were easy to prove, whereas we don't know other truths which are necessary (of the same sort of necessity) because they are difficult to prove or, more generally, they involve sophisticated reasoning we are unable to carry out: one can know from elementary school that $7 + 5 = 12$ without knowing that $x^n + y^n = z^n$ has no solutions in positive integers for $n > 2$. One can instantly see that $\varphi \supset \varphi$ is a logical truth without knowing that a long and complicated propositional tautology is.

Next, sometimes we think that φ without thinking that ψ, although φ entails ψ, because we lack some concept needed to grasp the proposition that ψ, and one cannot have such attitudes as knowledge, belief, or even supposition, towards a content one cannot grasp. One can, for instance, believe that φ without believing that $\varphi \wedge (\varphi \vee \psi)$ although the two are logically equivalent in classical and in various non-classical logics, for one lacks some concept which is needed in order to grasp ψ's content. Perhaps one cannot even see that φ if one lacks concepts required to grasp the proposition that φ, although one can see a situation in which φ (Barwise and Perry 1983; Williamson 2000):   one can see a situation in which Greta is playing Go, but one cannot see that Greta plays Go if one has no idea of what Go is.

Next, sometimes it's the intrinsic nature of the state that stands in the way of full closure, even for 'ideally astute logicians' (Dretske 1970, 1010) who, in addition, have the full repertoire of concepts at their disposal. Some propositional attitudes fail full closure in an obvious way: one desires or hopes that one's headache goes away, and this entails that one has a headache; but one doesn't desire or hope that one has a headache, even as an ideal logician (supposing such logicians can still have headaches).

Finally, other attitudes may fail full closure in a less obvious way. Full epistemic closure is at times contested also for accommodating interpretations of '$X\varphi$', e.g., not as saying that one knows φ, but, roughly, that one is in a position to know it bracketing contingent obstacles as well as cognitive and computational limitations. Some philosophers have seriously entertained the idea that perfect reasoners may know that they have hands, without being positioned to know that they are no recently envatted handless brains, although the having of hands entails that one is no handless brain in a vat.

What we need, is to be clear on what we are after in our modelling activities. We may want to idealize in one dimension, without this implying that we aim at giving a normative account for that dimension. It may turn out that the agents we model, idealized in certain respects, also happen to represent some sort of normative standard for those respects. But that's not the main reason why we idealize: we do it chiefly in order to have a simple and manageable setting to work with there, while we examine the effects of de-idealization in another, orthogonal dimension, which is the focus of our modelling aims. Idealization is for us just useful simplification here. For instance: we may want to represent agents with a limited conceptual repertoire, but whose relevant attitudes are perfectly closed with respect to any proposition graspable to them, to study how conceptual limitations work in a 'pure' setting, i.e., one in which limitations of other kinds don't get in the way. Or, we may want to represent agents who are deductively *and* conceptually unbounded, but whose epistemic position is defeasible because they operate on limited and potentially misleading, even when truthful, information. Or, we may want to grant such agents attitudes which are not fully closed, just because we think that's the intrinsic nature of the relevant attitudes.[4] As the story told in this book unfolds, the hope (though I may have failed to live up to it throughout all of the work) is that it is clear enough what kinds of idealizations and de-idealizations are in place in its various parts.

What's in the story? The following section gives a chapter-by-chapter overview. It defers to subsequent chapters for defenses of the main claims, but it's hopefully detailed enough for a first pass.

---

[4]Some examples of works that idealize in one dimension in order to better focus on another: Williamson (2000) highlights the opportunity of assuming logical omniscience to study the effects of limited powers of discrimination on agents, and advocates failure of introspective principles for such agents (they don't always know that they know). Holliday (2012) focuses on 'ideally astute logicians' à la Dretske, who 'know all logically valid principles and [...] believe all the logical consequences of what they believe' (92) to study possible failures of epistemic closure for them – thus, closure failures which are not due to deductive limitations of the modelled agents. Presenting awareness logics, Schipper (2015) proposes a full S5 modal logic for epistemic operators, thus representing fully positively and negatively introspective epistemic agents (not only are they logically omniscient: also, they always know both what they do and what they don't know), in order to investigate the consequences of limitations in conceptual awareness.

1.3   What's in the Book

The core idea is that the topic-sensitivity of *de dicto* thoughts pivots on the one of the propositions making for their contents. Now chapter 2 develops the view that propositions can be usefully seen as featuring (at least) two constituents: (1) truth conditions and (2) subject matter or topic. To say the same thing – to express the same content – sentences φ and ψ must coincide in both truth conditions and topic. The chapter calls *two-component* (2C) *semantics* the view that such constituents are not only usefully represented as distinct, but also, really irreducible to each other, in a sense to be clarified there. It would have been nice to call it 'two-dimensional semantics', in order to highlight the orthogonality of (1) and (2); but, alas, the name was already taken: see Schroeter (2021). Various subject-matter-sensitive accounts of propositional content are one-component (1C): they either reduce truth conditions to subject matter, or vice versa. However, there are 2C approaches on the market, too, e.g., those by Epstein (1981, 1993); Hawke (2016, 2018); Plebani and Spolaore (2021); and also one of the semantic proposals in Yablo (2014) counts as 2C.

That the two components are irreducible is, I believe, a stronger claim than what is needed to get a topic-sensitive logic of intentionality going. Hopefully, subsequent chapters will show that we have good theoretical reasons for representing the components as distinct, even if either is ultimately reducible to the other. We (myself and Peter Hawke, the co-author of chapter 2) try the stronger stance nonetheless. It pops up a few times in the book, that real irreducibility may be philosophically more satisfactory than useful pretence.

Anyway, our own 2C semantics agrees with various other subject-matter-sensitive semantics, also of the 1C kind, on a number of features of subject matters, which are introduced and discussed in chapter 2. These are important for the development of a topic-sensitive logic of intentionality. In several forms of such semantics, logically or necessarily equivalent sentences φ and ψ can differ in their propositional content when they are about different things, that is, they have different topics. The semantics are, thus, hyperintensional, making distinctions more fine-grained than what standard intensional (possible worlds) semantics allows. 'Equilateral triangles are equiangular' and 'Either Peter passed the exam, or not' differ in content even if they are true at the same possible worlds (all of them), for they are about different things: only one is about equilateral triangles and how they are like. Even sentences which are necessary of the same kind of necessity can so differ: '2 + 2 = 4' is not about equilateral triangles either.

Can't such distinctions be made just by resorting to some structured account of propositions (Soames 1985; King 1996; Chalmers 2011), whether Russellian structures of denotations, or structures of Fregean senses, or so? The structured proposition that 2 + 2 = 4 differs from the one that equilateral triangles are equiangular in various ways, e.g., by including a constituent (say, the number 2), which is not included in the other.

Well, in answer to this, let us start by remarking that our 2C semantics will agree with a number of other subject-matter-sensitive semantics, also of the 1C kind, in taking various sentential operators of propositional logic as *topic-transparent*: such operators add no subject matter of their own. We echo the Tractatian Wittgenstein's 'fundamental thought', that 'the "logical constants" do not represent' (4.0312). In particular, for negation:

> 4.0621 That, however, the signs 'p' and '$\sim p$' *can* say the same thing is important, for it shows that the sign '$\sim$' corresponds to nothing in reality. That negation occurs in a proposition, is no characteristic of its sense ($\sim\sim p = p$). [...]

> 5.44 [...] And if there was an object called '$\sim$', then '$\sim\sim p$' would have to say something other than 'p'. For the one proposition would then treat of $\sim$, the other would not. (Wittgenstein 1921/22)

The topic of $\neg\varphi$ is the same as that of φ. 'John is not tall' is exactly about what 'John is tall' is about, say: John's height. Additionally, conjunction and disjunction merge topics. The topic of $\varphi \wedge \psi$ is the same as the topic of $\varphi \vee \psi$, namely the fusion of the topic of φ and that of ψ. 'John is tall and handsome' and 'John is tall or handsome' are both about the same topic, say: the height and looks of John (Fine 2020, 136 makes a forceful case).

If these are constraints on the topics of the relevant propositions, they should correspond to constraints on the attitudes having such propositions as their contents: if, in order to think that John is tall, you have to think, say, about John's height, then that's exactly what you must think about in order to think that John is not tall. If, in order to think that John is tall and handsome, you have to think, say, about John's height and looks, then that's exactly what you must think about in order to think that John is tall or handsome.

Now this won't automatically work if we resort to structured propositions. Whatever their merits, one needs to fix something for them to deliver a plausible account of subject matter. The structured proposition that John is not tall differs from the one that John is tall by including *not* as a constituent. This won't work for aboutness – a point made in (Yablo 2014, 1): that proposition may be about John, John's height, how John is like, but surely it's not about *not*. And so it will be for the aboutness of intentional states having that proposition as their content: when you think that John isn't tall, you are not thinking about negation. (That doesn't make negation unthinkable, of course: you can think that negation is a one-place connective, for instance.) Similarly, when you think that John is tall and handsome, you are not thinking about *and*, although that, too, is a constituent of the structured proposition.

The topic-sensitive setting suggests a number of closure and non-closure properties for propositional attitudes, which are explored starting from chapter 3. When the topic of proposition P is x, and one thinks that P, one must be – stealing a couple of Yablovian metaphors – 'attentive to everything within x'; but one can be 'oblivious to matters lying outside of x' even when certain propositions having those (subject) matters as their topic are entailed by P: cf. Yablo (2014), p. 39.

Once one factors out forms of logical non-omniscience due only to certain cognitive and computational limitations (difficulties in parsing the syntax of sentences, the boundaries of people's reasoning and memory capacities), claims like the following seem plausible: one cannot know that Lisa is rich and happy without knowing that she's rich (Conjunction Elimination); for not only is the latter known proposition entailed by the former, but also, what it's about is *part of* what the former is about. One cannot imagine that Jack is short and thin without imagining that he is thin and short (Conjunction Commutation) when 'and' stands for Boolean or order-insensitive conjunction; for not only are the two logically equivalent, but also, they are about the same topic. One cannot believe that Mary is funny and that Mary is happy without believing that she's funny and happy (Conjunction Introduction); for not only is the latter proposition entailed by the former, but also, its topic is nothing but those of the former, taken together.[5] Such closure features of some intentional states, mandated by the mereology of the involved propositions' topics, have been labelled *immanent closure* by Steve Yablo (2014, 2017), and I will adopt the terminology.

The idea that thoughts are topic-sensitive and, because of this, 'closed under topicality' rather than fully closed under (classical, modal) logical consequence, is gaining popularity. It can also be found, e.g., in the works of authors like Yalcin (2016), and Hoek (2022), who focus on belief (both stress that beliefs are sensitive to *questions*; but, as we will see, there are tight connections between topics and questions). The topic sensitivity of thought naturally delivers, in fact, a number of *in*validities, too: when you think that John is tall, you needn't automatically be thinking that he's tall or handsome, even if $\varphi \vee \psi$ is just one single elementary inferential step away from φ and *even* if you are a deductively unbounded agent. You may not be thinking about John's handsomeness at all (your mental state isn't automatically Additive), for a number of reasons which will be explored in some detail in the book.

---

[5]If you have issues with this last, it may be due to the fact that you are taking the belief attitude as triggered by the passing of some intermediate probabilistic or degree-theoretic threshold, and you have doubts connected to Lottery- or Preface-Paradox-related considerations. Chapters 3 and 8 will deal with issues concerning thresholds, probabilities, and lotteries. Chapter 4 will say something on the Preface Paradox.

Also, even if you think that φ, and φ strictly implies ψ, that is, it just cannot happen (there is no possible world where) φ holds but ψ fails, you may think that φ without thinking that ψ, even if you are deductively unbounded, because strict implication fails to be topic-preserving. The topic of ψ may be *alien* to you: you may lack the conceptual resources to grasp it. And even if you are a deductively *and* conceptually unbounded agent, you may not be in such a favourable epistemic position with respect to ψ as you are with respect to φ, due to the topic addressed by ψ. Given certain empirical information, you are in a position to know that you have hands. There's no way you can be a handless brain in a vat if you have hands. But some epistemologists think that that very same information may not put you in a position to know you're no brain in a vat: your information, they may say, settles epistemic issues about everyday experience; it does not address the topic of far-fetched, sceptical scenarios.

Chapter 3 introduces a basic formal semantics for a family of Topic-Sensitive Intentional Modals (TSIMs, read 'ZIMMs'): modal operators representing attitude ascriptions, and which embed a topic-sensitivity constraint. The TSIM operators in focus in it, and in the following three chapters, are two-place, variably strict modals (we will see exactly what this means in due course), of the form '$X^\varphi\psi$' (generic reading: 'Given φ, one Xs (or, would X) that ψ', X expressing the relevant attitude ascription), with a topicality constraint linking φ and ψ. These are both non-monotonic, thanks to their variable strictness, and such that they fail full closure under logical consequence or entailment, thanks to their topic-sensitivity.

Variable strictness and topic-sensitivity differentiate the TSIMs explored in these chapters from the standard knowledge and belief operators à la Hintikka. Another feature brings back some similarity with the standard framework: starting from the basic semantics presented in chapter 3, one can add constraints on the accessibility relations or functions used in the truth conditions for the TSIMs. Such constraints validate different logical principles and entailments under which the operators are, or become, closed; and they also suggest different readings of the operators themselves. That's similar to how, in a standard modal setting, starting from

K (the basic normal modal logic), one gets stronger modal systems such as T, B, S4, S5, by adding (more or less controversial) constraints on the various accessibility relations; such constraints deliver certain validities, and make plausible certain interpretations for the corresponding operators.

Three (families of) constraints are explored in the three chapters following chapter 3, giving interpretations of the TSIMs as expressing, respectively: (i) Knowability Relative to Information (KRI; chapter 4); (ii) imagination as Reality-Oriented Mental Simulation (ROMS; chapter 5), carried out in suppositional thought; and (iii) hyperintensional conditional belief or (static) belief revision (with hints at a dynamic expansion; chapter 6).

In chapter 4, my co-author (Peter Hawke again) and I argue that the KRI setting ('$K^\varphi\psi$': 'Given total (empirical) information φ, one would be in a position to know ψ') has a lot to say on the big debate around epistemic closure, 'one of the most significant disputes in epistemology over the last forty years' (Kvanvig 2006, 256).

Roughly: given that one knows φ and (one knows that) φ entails ψ, is one in a position to know that ψ? Closure deniers (for instance, Dretske 1970; Nozick 1981; Lawlor 2005; Holliday 2015; Hawke 2016; Sharon and Spectre 2017) limit closure to address sceptical worries and defend fallibilism (in one typical formulation: one can sometimes know that φ even if one is not in an epistemic position to rule out all possible scenarios in which φ fails). Closure supporters (for instance, Williamson 2000; Hawthorne 2004; Roush 2010; Kripke 2011a) often stress that egregious violations of specific closure instances, e.g., of Conjunction Elimination, are unacceptable (how can one who knows $\varphi \wedge \psi$ fail to *be in a position* to know that φ?). In general, one needs to have sufficiently strong restricted closure principles to vindicate obvious ideas, such as that competent deduction from known premises, e.g., in mathematical reasoning, *must* preserve knowability.

Peter and I argue that topic-sensitivity allows KRI to invalidate controversial forms of closure while validating less controversial ones, escaping egregious violations. Also, in the KRI setting the variable strictness of the relevant TSIMs models a certain non-monotonicity of knowledge acquisition. Unlike the standard modal framework for epistemic logic, KRI accommodates plausible approaches to the *dogmatism paradox*, due to Kripke and Harman, whereby knowing agents seem to be immune to rational persuasion via information bringing in new evidence. The paradox, in essence, is that it appears rational to ignore countervailing evidence to what they know. We show that the paradox can be split into two sub-paradoxes: one is dealt with via non-monotonicity; the other is handled by the limitations of closure delivered by topic-sensitivity. In spite of being non-monotonic and topic-sensitive, KRI also satisfies principles capturing the idea that knowledge must be stable, as per the venerable Platonic view of *epistéme*.

In chapter 5, I address a sort of imagination in play in ROMS: the kind of suppositional exercise we engage in all the time, when we try to guess what will happen if such-and-so turns out to be the case in 'What if' questions ('What will I do if I can't pay my mortgage anymore?') or when, counterfactually, we want to ascertain responsibilities ('Would he have managed to hit the brakes in time, if he had not been drinking before driving?'). In ROMS clothing, TSIMs ('$I^\varphi\psi$': 'Supposing φ, one imagines that ψ') are shown to model interesting features of such imaginative and suppositional thought.

The starting point of the chapter is a puzzle: given that imagination is anarchic and arbitrary (in ways belief, typically, is not), how can it have epistemic value? How can it give us knowledge of reality, if it is an unbounded departure from reality? The puzzle is addressed by distinguishing voluntary and involuntary aspects of ROMS. A number of plausible features of ROMS, taken from research in cognitive psychology and the philosophy of mind, are then listed; and it is shown that our topic-sensitive semantics can model them.

Also, the chapter considers the addition of a constraint on the semantics, whose effect is to validate a principle of 'equivalence in imagination', which limits the hyperintensional anarchy of imagination and strengthens its logic. Equivalence in imagination is a sort of cognitive equivalence: φ and ψ are equivalent in imagination for one, when they play the same role in one's cognitive life: whatever one infers, concludes, finds plausible, etc., supposing either, one does supposing the other, an idea I got from Levin Hornischer (2017).

In chapter 6, TSIMs are shown to help with conditional belief and (dispositions to) belief revision ('$B^{\varphi}\psi$': 'One believes ψ, conditional on φ'; or 'If one were to learn that φ, one would believe that ψ was the case'). The starting point is, again, hyperintensionality: it seems that we can believe different things conditional on necessary or logical equivalents which differ in topic. The things we believe, conditional on the proposition that Socrates exists, can greatly differ from those we believe, conditional on the proposition that Socrates' singleton, {Socrates}, exists. Additionally, we don't come to trivially believe everything just because we are, on occasion, exposed to inconsistent information. This is not straightforwardly represented either in the AGM framework for belief revision, or in various epistemic-doxastic logics recapturing AGM in a modal setting.

The addition to the basic semantics here consists in ordering the possible worlds essentially as in the mainstream Lewis-Stalnaker semantics for conditional logic – except that the (pre)ordering doesn't represent objective similarity, rather subjective plausibility, as in Grove (1988): the closer a world, the more plausible the scenario it represents for the believing agent. This setting has been used to model dispositions to revise beliefs, or belief entrenchment, in doxastic logic and Dynamic Epistemic Logic (DEL): Van Ditmarsch et al. (2008); Van Benthem and Smets (2015).

Besides exploring TSIM conditional belief or (static) belief revision operators, the chapter also hints at how to develop the topic-sensitive setting in a properly dynamic fashion via operators of the form '$[*\varphi]\psi$' ('After belief revision by φ, ψ holds), interpreted as model-transformers as in DEL. In particular, once topic-sensitivity is taken on board, we can have a dynamics involving the topics themselves: we can model how one comes to grasp new subject matters by expanding one's conceptual repertoire – an idea I have developed together with Aybüke Özgün.

Chapter 7 is joint work with Aybüke. Here we leave two-place TSIMs behind and focus on one-place TSIMs expressing two sorts of belief, in order to tackle typical forms of well-known *framing effects* (Kahneman and Tversky 1984). These, too, concern agents who can have different attitudes towards logically or necessarily equivalent propositions. Framing is known to have momentous psychological and social consequences. Unlike Econs, the fully consistent agents of classical economic theory who well-order their preferences and maximize expected utility, Humans can be 'framed': nudged into believing different things depending on how equivalent options are presented to them (Thaler and Sunstein 2008).

Typical framing effects seem to be a more specific phenomenon than the general hyperintensionality of belief (revision) modelled in the previous chapter: framed believers, of the kind studied in cognitive psychology, choice theory, and behavioural economics, can have different attitudes towards co-intensional contents even when they (the agents) are on top of all the relevant subject matters and, in addition, they can even be, in some sense, aware that those contents are equivalent, although in another sense they certainly are not. Besides the topic-sensitivity of belief ascriptions, this application of TSIM theory relies on modelling the structural distinction, taken from cognitive psychology, between beliefs activated in working memory (WM) and beliefs left inactive in long-term memory (LTM). Framed agents can have the belief that patients should get surgery with a 90% one-month survival rate activated in their working memory, without having the intensionally equivalent belief that patients should get surgery with a 10% first-month mortality there. However, such agents can have all the relevant information as well as the concept *mortality* in their (declarative) LTM. Calling beliefs activated in WM *active* and beliefs left asleep in LTM *passive*, this chapter provides a logic of topic-sensitive active and passive belief for framed agents: a belief is active when it is available in WM to perform cognitive tasks with it. It is passive when it is stored, or encoded, in the agent's LTM knowledge base, and left inactive there.

Up to chapter 7 included, belief and the other attitudes

modelled by the TSIMs are taken as all-or-nothing and non-probabilistic. In this respect, TSIM theory is still in line with standard epistemic logic. However, other approaches to the logic of intentional states are quantitative, graded and, typically, probabilistic (Halpern 2005). In particular, belief is often understood in terms of, or at least linked to, degrees of confidence taken as subjective probabilities, and Bayesianism dominates the literature on belief revision.

The final chapter of the book begins to scratch the surface of the connections between topicality and probabilities. One popular philosophical approach to non-monotonic indicative conditionals (for instance, Adams 1975; Edgington 1995; Bennett 2003) understands these probabilistically, too, and in strict connection to conditional belief (as strict as allowed by Lewis (1976)'s notorious triviality results). The idea has become mainstream also in the so-called 'New Paradigm' in the psychology of reasoning (Evans et al. 2003; Evans and Over 2004; Oaksford and Chater 2010), where probabilistic, non-monotonic, and conditional reasoning are tightly related.

Now the two-place TSIMs of chapters 3 to 6 are a sort of non-monotonic, conditional-like operator. This led Aybüke and me to consider how topic-sensitivity may relate to probabilities in the abstract setting of conditionality. And so, in chapter 8 we provide a semantic framework that, leaving possible worlds behind, spells out acceptability conditions for a topic-sensitive indicative conditional which adds a topicality component to a treatment in terms of conditional probabilities (we use Popper functions: conditional probabilities are primitive, not defined via the Ratio Formula in terms of unconditional ones).

We remain neutral on the tricky issue of whether indicatives express propositions and can generally have truth values. We take as our starting point Adams' Thesis (Adams 1966, 1998): the idea that the acceptability of a simple indicative (one with no indicatives embedded in its antecedent or consequent) equals the corresponding conditional probability. The Thesis enjoys considerable popularity in philosophy but, we argue, it's false and refuted by recent empirical results, which we summarize.

In our setting, a simple indicative '$\varphi \to \psi$' is to some extent linked to the idea that ψ follows from φ and background, unstated assumptions connected to φ ('BA_φ'). $\varphi \rightarrow \psi$ is acceptable to the extent that (1) $p(\psi|\varphi)$ (the probability of ψ, conditional on φ) is high, provided (2) ψ is on-topic with respect to φ and BA_φ. This gives a kind of relevant conditional, unacceptable when a topicality connection between antecedent (and background assumptions) and consequent is missing. We show that the validities and invalidities in the probabilistic logic of such a conditional are both theoretically desirable, and in line with experimental results on how people reason with conditionals.

1.4   ... And More to Come

I said at the outset that this book begins to explore an idea. I meant it. Various views it experiments with are new and rather tentative, various discussions in it are to some extent inconclusive, and I'll be surprised if problems I haven't foreseen (it often happens to me!) don't pop up soon. I am happy, however, that the idea of topic-sensitive logics of thought is already being picked up by a number of logicians and philosophers, who have paid attention to papers on which this book is based and have variously criticized and/or developed them.

Ale Giordani (2019) has axiomatized the logic of imagination of chapter 5, coming up with a system that's sound and complete with respect to the semantics presented there. Additionally, some have focused on better representing the agentive nature of imagination as ROMS that chapter deals with: Ale again and Ilaria Canavotto, with some help from me (Canavotto et al. 2020), have developed a fine-grained approach to voluntary imagination combining my ideas with a dynamic logic based on action types, and with tools from STIT logics (Segerberg et al. 2020). The idea of mixing my approach with STIT has also occurred to Heinrich Wansing, who has worked on it with Chris Badura: see Badura and Wansing (2021). Meanwhile, Chris has already upgraded the ROMS propositional setting, coming up with a sophisticated account of topicality and topic-inclusion for

a first-order language (Badura 2021b). Pierre Saint-Germier
(2020) has been exploring a non-2C approach where topic-
sensitive truthmaker semantics à la Fine is employed to give
a hyperintensional logic of imagination as ROMS, which
may do better than my attempt from chapter 5. Aybüke
again has provided a dynamic, topic-sensitive belief revision
operator and a sound and complete logic for it, with some
help from me (Özgün and Berto 2020), as per the results
mentioned at the end of chapter 6. She is also working with
my St Andrews colleague Aaron 'King of Mereology' Cotnoir,
exploring mereotopological developments of the very idea
of topic-inclusion (I'll say something on this at the end of
chapter 5).

Overall, topic-sensitive logics of intentionality make for a
nice new territory, and I hope more and more people get
interested in exploring it.

1.5   Chapter Summary

The logic of thought, understood as the logic of propositional
or *de dicto* intentional mental states, such as knowing,
believing, supposing, should be topic-sensitive. What ψs one
is to think as a consequence of thinking that φ should depend
on the propositions that φ and that ψ, which make for the
contents of such thoughts. And this book defends the idea
that propositions can be seen as individuated, not just by
the sets of worlds at which they are true, but also by what
they are about: their topic, or subject matter. The topic-
sensitive operators used to express attitude ascriptions, to
be explored in the book, fail full closure: given one such
operator X, one's Xing that φ does not imply that one Xs
all of φ's entailments. On the other hand, such operators
are not logically anarchic. Failure of full closure connects to
the problem of logical omniscience, and we have seen that
non-omniscience has distinct, independent sources. We have
then been through a summary of the contents of the coming
chapters, introducing their main ideas, and we have pointed
at recent or in-progress developments of such ideas.

2

Two-Component Semantics

Co-authored with Peter Hawke

The logic of thought must rest on an account of propositions, if propositions are the contents of *de dicto* intentional states. In this chapter, we provide such an account. Its most distinctive trait is the idea that propositions can be usefully represented as featuring (at least) two constituents: (1) truth conditions and (2) subject matter or topic. We also argue that one may have reasons for taking the two components (1) and (2) as irreducible to each other, in a sense to be clarified. However, one need not accept the irreducibility proposed here to appreciate the logics presented in the following chapters. There, topic-sensitive intentional modal operators, or TSIMs, get a two-component semantic interpretation in terms of possible worlds and topics. One may find such a twofold setting theoretically useful, for instance, because it delivers certain desired logical validities and invalidities for the operators, although the two-component representation ultimately involves some redundancy. The current chapter will still have the useful role of familiarizing its reader with subject matters in a general setting.

This said, we try the stronger stance here. One reason for doing so is that real irreducibility – the distinction between (1) and (2) 'carving at the natural joints', to adopt the usual metaphor – may be more satisfactory philosophically, although some logicians may not care about philosophical satisfaction, take a more instrumentalist stance, and just look at what the logics deliver.

Topics of Thought: The Logic of Knowledge, Belief, Imagination.
Francesco Berto, Oxford University Press. © Francesco Berto 2022.
DOI: 10.1093/oso/9780192857491.003.0002

We start by presenting the general idea of a topic-sensitive view of propositional content, and introduce the distinction between one-component (1C) and two-component (2C) accounts, in Section 2.1. We do it via a thesis, which we name after Yablo, and we distinguish a Full and a Weak formulation of it.

In Section 2.2, we discuss how a mereology of topics may work (Subsection 2.2.1) and we list a number of constraints any topic-sensitive account of propositional content should, we think, obey. One series of constraints concerns the topic-transparency of some logical vocabulary (Subsection 2.2.2). Another one concerns there being topic-diverging necessities and co-necessities: sentences which are, respectively, necessary (of the same kind of necessity), or necessarily equivalent, but which express different propositions because they are about different things (Subsection 2.2.3).

In Subsection 2.2.4, we argue that topic-sensitivity, in spite of playing a key role in the logics of intentional states to be presented in the following chapters, may be kept distinct from other phenomena concerning intentional contexts and, in particular, Fregean puzzles, which are often dealt with by invoking representational guises, or modes of presentation that would be in play in attitude ascriptions.

In Section 2.3, we speak of 1C semantics in a general setting. We come up with a *prima facie* case, to which 1C theorists should react, for truth conditions and topics being irreducible to each other (Subsection 2.3.1). 1C semantics has it that either truth conditions are reducible to subject matter, or vice versa. We argue that the first disjunct stands in tension with Transparency (Subsection 2.3.2); and that the second may not account for topic-diverging (co-)necessities very easily (Subsection 2.3.3).

In Section 2.4, we introduce a formal 2C semantics which delivers verdicts on same-saying, that is, on which sentences express the same proposition taken as composed of a truth set and a topic. We show that such semantics complies with the constraints presented in Section 2.2 in a straightforward way.

In Section 2.5, we examine the prospects for the Full Yablo's Thesis. We discuss two objections to it: we call them the Objection from Mathematical and Logical Ignorance (Subsection 2.5.1), and the Objection from Classicality (Subsection 2.5.2). Our response to these, and especially the latter, is rather tentative and points at deep issues that will resurface in later chapters.

2.1   Yablo's Thesis, 1C, 2C

Declarative sentences enable us to say *true* things *about* all sorts of *topics*. One says: 'Jane is a lawyer'. One thereby communicates something about Jane's profession, what Jane does, and, more generally, Jane. What one says is true just in case Jane's profession is or includes being a lawyer. One addresses certain topics and says such-and-so about them.

One may think that it's fruitful, then, to model a proposition as a pair:

$$P = \langle \mathsf{C}_P, \mathsf{T}_P \rangle$$

C_P gives the set of Circumstances under which P is true (its truth conditions), and T_P gives the Topic of P (its subject matter: what it's about). We talk generically of 'circumstances' for the following reason. We will phrase our own 2C semantics below, as well as the formal semantics of the coming chapters, using classical, maximally consistent possible worlds; thus, these will be our working circumstances for the rest of the book.[1] But in this chapter, we will discuss

---

[1] The default will be that the possible worlds used in our semantics throughout the book, taken together, represent so-called absolute or unrestricted possibility and necessity (it will be flagged when this is not quite the case). These are at times labelled as 'metaphysical', especially after Kripke: metaphysical necessity is 'necessity in the highest degree', says Kripke (1980), 99. Williamson (2016b) calls metaphysical possibility the 'maximal objective modality' (459). And Stalnaker (2003) says: 'we can agree with Frank Jackson, David Chalmers, Saul Kripke, David Lewis, and most others who allow themselves to talk about possible worlds at all, that metaphysical necessity is necessity in the widest sense' (203). So 'we begin with the idea of the totality of possible worlds across which all of the genuine possibilites (and no impossibilities) are represented' (Divers 2002, 5), and claim that to be unrestrictedly or metaphysically necessary is to hold throughout all such totality. What sorts of necessities count as absolute or unrestricted is, of course, controversial. Mathematical truths like '2 + 2 =

other approaches whose truth- (and falsity-) supporting items will be different from classical possible worlds. If P is the content of φ, one may call $\langle C_P, T_P \rangle$ the *thick* proposition associated with φ; C_P, the *thin* proposition (Yablo 2014, 3.3, uses this terminology).

Semantics has traditionally focused on truth conditions, given via (sets of) possible worlds, or situations à la Barwise and Perry (1983), or in other ways yet. These are often taken as primary (Heim and Kratzer 1997, ch. 1), whereas topics are often judged irredeemably vague or elusive (Ryle 1933; Perry 1989). Nevertheless, philosophers (Putnam 1958; Goodman 1961; Lewis 1988a), linguists (Roberts 2011), and logicians (Fine 1986) have not ignored topicality. As mentioned in chapter 1, work in this area has been burgeoning lately (Gemes 1994, 1997; Humberstone 2008; Fine 2016a, 2017; Felka 2018; Moltmann 2018; Plebani 2020). And rightly so, because topics are crucial in a theory of *partial content*, which, as we will see in the coming chapters, is in its turn crucial for a theory of the logic and semantics of attitude ascriptions.

Some examples starring Jane:

1. Jane is a lawyer and an expert footballer.

2. Jane is an expert footballer and a lawyer.

3. Jane is an expert footballer.

4. Jane is a footballer.

The natural view we may take as our starting point, is that (1) and (2) are true/false exactly in the same circumstances, (1) entails (3), as does (2), and (3) entails (4). This needn't

---

4' or logical truths like '$\varphi \supset \varphi$' are plausible candidates. Thus, a circumstance in which 2 and 2 don't add up to 4, or $\varphi \supset \varphi$ fails, would make neither for a possible world nor for a part thereof: things *just couldn't* be or have been like that. If there are 'analytic necessities' at all, these may fit in as well: given that 'If someone is horribly late, then someone is late' counts as one, a circumstance where one is horribly late but not late is not a possible world, nor a part thereof. People also understand 'metaphysical necessity' in a narrower sense, as absolute necessity which is only knowable *a posteriori*. 'Water is H2O' and 'Socrates is human' would be (for many essentialists) good examples. These are set apart from those other examples of metaphysical (in the broad sense) necessities, if logical and mathematical (and analytical, if such there be) truths are *a priori*.

be taken as reflecting perfectly general facts about adverbial modifiers or 'and' in natural language, of course: 'Jane is usually late' does not entail 'Jane is late'. Though debatable, it is plausible that in various conversational contexts 'Maria went to the hospital and got sick' is not entailed by 'Maria got sick and went to the hospital'. In a prominent class of cases including natural conversational contexts for (1) and (2), however, conjunction is not order-sensitive.

One may also think, naturally enough, that (1) and (2) have the same content (Ann claims: 'Jane is a lawyer and an expert footballer'; Bob claims: 'Jane is an expert footballer and a lawyer': Ann and Bob have said the same thing. Had Bob replied to Ann: 'No, actually Jane is an expert footballer and a lawyer', we would resort to pragmatics and conversational clues: maybe Bob wanted to emphasize that playing football is what Jane really loves, or to implicate that she is not an expert lawyer); that (3) expresses a proper part of that content (Carlos claims: 'Jane is an expert footballer'; what he says has already been said by Ann and Bob – who also said more: that Jane is a lawyer, too); and that (4) expresses a proper part of the content of (3) (Dave says: 'Jane is a footballer': what he says has already been said by Carlos – who also said more: that she is expert *qua* footballer).

What is it for φ to say whatever ψ says, that is, for the content of φ to contain that of ψ as a part (let's abbreviate this as: '$\varphi \trianglerighteq \psi$')? Here's a proposal, with due credit to a very fine philosopher:

(Yablo's Thesis) $\varphi \trianglerighteq \psi$ iff (1) φ entails ψ and (2) ψ is about topic x only if φ is about x.

'Content-inclusion is implication plus subject-matter inclusion' (Yablo 2014, 15). One may call the conjunction of (1) and (2) *Parry Implication*; see Parry (1968, 1989). This gives an account of same-saying as two-way containment: φ and ψ say the same ('$\varphi \trianglelefteq\trianglerighteq \psi$') just in case they are both mutually entailing and topic-equivalent.

Only the left-to-right direction of Yablo's Thesis (call it *Weak Yablo*) is needed for many of our purposes throughout this book, and in particular, to put to work our topic-sensitive logics of thought in subsequent chapters. We can run with the

Full Yablo's thesis for now, but we will discuss whether one is better off accepting Weak rather than Full Yablo in Section 2.5.

Are facts about topicality reducible to facts about truth conditions? Are facts about truth conditions reducible to facts about topicality? An affirmative answer to either question gives a *one-component* (1C) view, which implies a certain redundancy in saying that φ and ψ are both mutually entailing and topic-equivalent. 1C can be taken as coming with a disjunctive claim: either topics are a function of truth conditions, or truth conditions are a function of topics, where 'function' is plainly understood as a procedure to extract, for any proposition P, either component from the other. We will give a more roundabout characterization, which will turn out to be useful for our purposes:

(1C) Either, for every proposition P, knowing C_P is *a priori* sufficient for knowing T_P, or, for every proposition P, knowing T_P is *a priori* sufficient for knowing C_P.

Knowing X is *a priori* sufficient for knowing Y when one who knows X is in a position to know Y *a priori*. Roughly: if the X-knower were freed from contingent obstacles and cognitive (computational, temporal, etc.) limitations, the X-knower would get to know Y with no need for help from further empirical information, just by thinking hard about the *a priori* consequences of what the X-knower knows.[2]

We can still abbreviate 'Knowing X is *a priori* sufficient for knowing Y' using functional notation: there's an f such that $Y = f(X)$, where f is a function encoding a procedure to extract Y from X *a priori*. 1C is, then, the claim that either there is an f such that $T_P = f(C_P)$ for every proposition P, or there is an f such that $C_P = f(T_P)$ for every proposition P.

2C semantics implies a denial of the 1C claim above. We call a compositional sentential semantics – a recursive procedure to assign contents to sentences – a *one-component semantics*

---

[2]The literature around *being in a position to know* has been burgeoning especially since Williamson (2000) made a key use of the notion in his 'knowledge first' approach to epistemology. One notable paper on the topic is Hawthorne and Yli-Vakkuri (2020).

if it entails 1C. A *two-component semantics* entails 2C. Or, more precisely: suppose the job of a formal semantics for a language $\mathcal{L}$ is to define a class of models for $\mathcal{L}$, with each model assigning a proposition – truth conditions and a topic – to each sentence in $\mathcal{L}$. Each model supplies a class of propositions, with each proposition P associated with truth conditions C_P and a topic T_P. Then a formal semantics is a one-component semantics just in case for every model $\mathcal{M}$, there exists a function f such that either $T_P = f(C_P)$ for every proposition P relative to $\mathcal{M}$, or $C_P = f(T_P)$ for every proposition P relative to $\mathcal{M}$. A formal semantics is a two-component semantics just in case there exists a model $\mathcal{M}$ for which there are two propositions that have the same truth conditions but differ in topic, and two propositions that have the same topic but differ in truth conditions. A 2C-semantics can allow all kinds of important connections between truth conditions and topics. But according to 2C, facts about either cannot be reduced to facts about the other in the sense described above.

Various topic-sensitive semantics can be interpreted as 1C. In Lewis (1988a,b)'s approach, the circumstances in C_P are possible worlds. Topics are world-partitions dividing logical space into ways a certain subject can be. Topics are determined by truth conditions: P is about x when x refines the binary partition composed of P's truth set (the set of worlds at which it is true) and its complement. Thus, the procedure f to extract T_P from C_P can be understood as either the identity function – identifying the subject matter of P with the aforemenoned binary partition – or the function mapping the binary partition to the set of its refinements.

In one of the settings considered in Yablo (2014), T_P is a pair: a set of truthmakers (subject matter P, say) and a set of falsemakers (subject anti-matter $\overline{\mathsf{P}}$). Truthmakers and falsemakers are sets of worlds, together forming a way of dividing logical space: the ways in which P can be true, plus the ways in which it can be false. (Yablo ends up favouring divisions, which admit overlap, over Lewisian partitions, for the sake of generality. This is immaterial for our discussion.) Then, the truth set for P is the union of its truthmakers. If T_P is taken to be an ordered pair $\langle \mathsf{P}, \overline{\mathsf{P}} \rangle$, then P's topic determines its truth conditions: the procedure f to extract the latter from the former has it that $C_P = \bigcup P$.

In the approach of Fine (2016b, 2017), C_P can be identified with a pair of sets of states, which, unlike classical possible worlds, may be neither possible nor maximal: a set of 'exact verifiers' and a set of 'exact falsifiers'. Topic is determined by truth(making) conditions: f has it that T_P is the fusion of every state in the union of the sets in C_P.

There are 2C approaches on the market, too. The 'relatedness logic' by Epstein (1981, 1993) is the earliest we know of. The considered view of Yablo (2014), 2.8, is that T_P is best modelled as an *un*ordered pair $\{P, \overline{P}\}$. This looks like a 2C account: P and Q may have the same topic $\{X, Y\}$ but differ in truth conditions, with X giving the truthmakers for P, Y giving those for Q. Further 2C approaches have been elaborated by one of us (Hawke 2016, 2018). Also, Plebani and Spolaore (2021) were mentioned in chapter 1: they modify Lewis' partition-based account into a 2C version in a simple and intuitive way. Furthermore, a Finean might want to *explicate* 'exactness' in terms of a prior, independent account of subject matter. This may fit more comfortably under the 2C banner.

2.2   Topicology

We will use a simple formal language $\mathcal{L}$ for our topicological purposes. $\mathcal{L}$ has singular terms and two-place predicates, negation $\neg$, conjunction $\wedge$, disjunction $\vee$, the box of necessity $\Box$, meta-semantic operators $\unrhd$ and $\approx$, round parentheses as auxiliary symbols $(,)$. We use $\varphi, \psi, \chi, \ldots$, as metavariables for formulas of $\mathcal{L}$. If a and b are singular terms, R a two-place predicate, aRb is an atomic formula. The well-formed formulas are the atomic formulas and, if φ and ψ are well-formed formulas, so are the following:

$$\neg\varphi \mid \Box\varphi \mid (\varphi \wedge \psi) \mid (\varphi \vee \psi) \mid (\varphi \unrhd \psi) \mid (\varphi \approx \psi)$$

Outermost brackets are usually omitted. We identify $\mathcal{L}$ with the set of its well-formed formulas. Read '$\varphi \unrhd \psi$' as saying that the content of ψ is part of that of φ; let $\varphi \trianglelefteq\trianglerighteq \psi := (\varphi \trianglerighteq \psi) \wedge (\psi \trianglerighteq \varphi)$; read '$\varphi \approx \psi$' as saying that φ and ψ have the same topic.

A full-fledged 2C formal semantics for $\mathcal{L}$ is coming in Section 2.4. We now use bits of $\mathcal{L}$-notation to formulate and discuss general constraints on a good account of topics. We use x, y, z $(x_1, x_2, \ldots)$, for topics; $w, w_1, w_2, \ldots$, for worlds; $s, s_1, s_2, \ldots$, for states of affairs. States may be partial, or even impossible (we'll get back to this), if one follows Fine (2016b); one might then take worlds as maximal possible states. We use $P, Q, R, \ldots$, for arbitrary propositions; C_P for the truth conditions for P, given by the set of Circumstances in which P is true/false; and T_P for the Topic of P. A semantics will assign a (thick) proposition $[\varphi]$ to each well-formed sentence φ of $\mathcal{L}$. We use $\lfloor\varphi\rfloor$ for the truth conditions for $[\varphi]$, $\mathsf{C}_{[\varphi]}$, $\lceil\varphi\rceil$ for its topic, $\mathsf{T}_{[\varphi]}$.

What *are* topics? (If you've come this far in the book, you must have had that question in mind for a while!) To begin with, topics are often taken as being *of* interpreted sentences in discourse: the topic assigned to φ is what φ is about in a given conversational context (Yablo 2014, ix).

Next, on the one hand, topics are naturally linked to questions or issues in focus (Lewis 1988a,b; Roberts 2012): 'Our topic today is whether psychology is a mature science' maps to 'Is psychology a mature science?'. Topics needn't be framed as questions ('Our topic is the number of stars'); but there always seems to be a question in the vicinity ('How many stars are there?'). This motivated Lewis' partition-based approach: a question with a single good answer partitions logical space into equivalence classes. Two worlds end up in the same cell when they agree on the answer: all zero-star worlds, all one-star worlds, etc.

On the other hand, any old sort of thing seems to be able to serve, in some sense, as a topic: 'Our topic today is George Bush'; 'Our topic in this course is recursive functions'. Some worldly approaches to subject matter are more object-oriented, some are more fact- or states-of-affairs oriented: Hawke (2018) gives a critical overview; see also Yablo (2014), ch. 2. Reconciling them all can be tricky: the connection to questions pulls towards identifying topics with sets of distinctions, or issues under discussion. But the ease with which objects or states of affairs can be corralled to serve as topics pulls towards identifying these with parts of (concrete) reality.

Here's a way *not* to take a stance, for our logico-semantic purposes: given thing a, associate it with a topic, **a** (in this chapter, we use boldfaced fonts to talk about topics from now on), leaving it open whether **a** just is a, or the set of ways things can be with respect to a, or the bunch of states involving a as a constituent, or something else yet. What we care about, is not so much what kinds of things topics could be, metaphysically speaking, but what constraints they should obey for our logical and semantic purposes. Let's start looking into this.

2.2.1   The Mereology of Topics

The space of topics, whatever these are, must display some mereological structure (Yablo 2014; Fine 2016b, 2017): topics can have proper parts; distinct topics may have common, overlapping parts; one topic may be included in another in that every part of the former is also a part of the latter. **Mathematics** includes **arithmetic**. **Mathematics** and **philosophy** overlap, having (certain parts of) **logic** as a common part. **Jane's profession** includes the topic introduced by the question, 'Is Jane a lawyer?' – and in turn is included in **Jane**: to talk about whether Jane is a lawyer is to talk both about her and, more specifically, her profession.

We thus need *topic fusion*, $\oplus$, an operation merging topics x and y into a topic $x \oplus y$. For simplicity, we assume that any two topics can be fused, that is, fusion is unrestricted. $\oplus$ generates a partial order, *topic inclusion* or *topic parthood*, $\leq$, defined the usual way, $x \leq y := x \oplus y = y$, i.e., x is part of y when the fusion of x and y just is y. This kind of talk will pop up several times throughout this book.

How does this idea of a mereology of topics relate to what φ is about? Here are two candidate accounts. On the first one, φ is about x just in case $x \leq \lceil \varphi \rceil$. Then consider $\varphi =$ '$2 + 2 = 4$'. This can be, say, about **2** (which needn't be the same as the number 2), or **what 2 and 2 add to**; more broadly, it can be about **the natural numbers**. '$2 + 2 = 4$'

may be an appropriate reaction to 'Tell me about the simplest feature of the number 2 you can think of'; or to 'What do 2 and 2 add to?'; or to 'What's a basic truth concerning the natural numbers?'. This requires **2**, **what 2 and 2 add to**, and **the natural numbers** to be part of $\lceil\varphi\rceil$. But **3** and **what 3 and 3 add to** are part of **the natural numbers**. It follows, egregiously, that '2 + 2 = 4' is also about **3** and **what 3 and 3 add to**.

Here's a better account, which we will follow (Fine 2016b): φ is about x just in case $\lceil\varphi\rceil$ and x have a common part. φ is entirely about x just in case $\lceil\varphi\rceil \leq x$, and φ is partly about x just in case φ is about x but not entirely about x. This works better. Take '2 is even'. This is entirely about **the natural numbers**, thus: $\lceil 2$ is even$\rceil \leq$ **the natural numbers**. It doesn't follow that '2 is even' is about **3**: **3** is a proper part of **the natural numbers** that needn't overlap with $\lceil\varphi\rceil$.

We'll assume that, if φ is a simple predication, its topic can be expressed with 'whether φ'. ('Let's talk about whether 2 is even' is an invitation to discuss the topic of '2 is even'.) Consider:

5. 2 is even.

6. 2 is transcendental.

7. 4 is even.

Since (5) and (6) are about **2**, their topics overlap with **2**. Suppose the same part of **2** overlaps with their topics. Thus, these have a common part. So, (6) is about the topic of (5). So, '2 is transcendental' is about **whether 2 is even**. This seems wrong. Thus, the part of **2** that overlaps with the topic of (5) must be distinct from the part overlapping with the topic of (6). In particular: **2** is not contained in the topic of every claim mentioning 2. We can generalize.

Suppose (5) has topic $x \oplus y$, where only y is part of **2**. A plausible candidate for x would be the topic **whether something is even**, assuming 'Something is even' is not about **2** (which is debatable). *Mutatis mutandis*, x is part of the topic of (7). It follows that '2 is even' is about **whether 4 is even**. This seems wrong. In the absence of other plausible candidates for x, we conclude: (5) is entirely about **2**.

'$2 + 2 = 4$' and '$4 + 2 = 2$' share subject matter: both are about **2** and **4**. But their topic is not the same, cf. (Yablo 2014, sect. 2.1): only the first is about whether 2 and 2 add to 4. If the topic of conversation is whether 2 and 2 add to 4, it seems off-topic to start talking of what 4 and 2 add to. Generalizing, we should *reject*:

(Constituent Equivalence) $aRb \approx bRa$

This shows a certain irreducibility of the subject matters of whole sentences to the semantic values of their subsentential constituents. (We'll say more on this in Section 5.5.) It's not only what things one talks about that matters, but also what one says about them. To pick the Yablovian example: 'Man bites dog' and 'Dog bites man' refer to the same things (man, dog, perhaps biting), but say different things about them: the former has a more interesting subject matter than the latter (Yablo 2014, p. 24). Kit Fine (2020) agrees.

2.2.2   Transparency

The topics of logical complexes formed using negation, conjunction, and disjunction, should obey this recursion.

(Negation Transparency) $\lceil \neg \varphi \rceil = \lceil \varphi \rceil$

(Junctive Transparency) $\lceil \varphi \wedge \psi \rceil = \lceil \varphi \vee \psi \rceil = \lceil \varphi \rceil \oplus \lceil \psi \rceil$

Such connectives should be, that is, *topic-transparent*: they should add no subject matter of their own to the sentences they feature in. Several topic-sensitive semantic approaches, including those of Perry (1989); Epstein (1993); Beall (2016); Fine (2020), broadly agree on this. 'Jane is not a lawyer' is exactly about what 'Jane is a lawyer' is about. 'John is tall and handsome' and 'John is tall or handsome' are both about the same topic: the height and looks of John.

There is no obvious candidate for a topic that is systematically introduced by a negation, a conjunction, or a disjunction. What subject matter might 'Jane is not a lawyer' add to 'Jane is a lawyer'? 'Jane is not a lawyer' doesn't speak about **negation** (not that negation is ineffable: 'Negation is a logical connective' is about **negation**); nor is it about the larger topic **logic**. Nor can the new subject matter be understood as a new question, distinction, or issue: the minimally pertinent question 'Is Jane not a lawyer?' is, presumably, topically equivalent to 'Is Jane a lawyer?'. Similarly for 'Jane is a lawyer and Jane is a lawyer' and 'Jane is a lawyer or Jane is a lawyer'.

Besides, it is hard to come up with a discourse context where 'Jane is a lawyer' is on-topic, but 'Jane is not a lawyer' would be off-topic, or vice versa. Either seems on-topic with respect to the obvious topics the other is about: **whether Jane is a *lawyer*, Jane's profession, Jane**, etc. One easily imagines contexts where only one is *informative*. But this kind of irrelevance is easily distinguished from being off-topic: if we're discussing **Jane** and Ann says 'Jane is a lawyer', Bob's subsequent utterance of 'Jane is a lawyer' is uninformative, not off-topic; if Carl says '*e* is transcendental' then Carl has said something off-topic, whether or not the interlocutors commonly assume that *e* is transcendental.

That P and not-P have the same subject matter does not, of course, imply that **P** and **not-P** are the same topic. It is, for instance, plausible that 'not-P is possible' and 'It would be a disaster if not-P were the case' are about **not-P**, but not about **P**. Further, the scope of a negation should (as always) be carefully attended to. For instance, Negation Transparency is compatible with thinking that one can talk about who might be at the party without talking about who might not be at the party. In such a conversation,'Jane might not be there' is off-topic despite 'Jane might be there' being on-topic. Not so, according to Negation Transparency, for the negation of the latter: 'Jane cannot be there'.

It's also hard to think of a discourse context where 'Jane is a lawyer' is on-topic, but 'Jane is a lawyer and Jane is a lawyer' is not: if the former is on-topic, the latter is objectionably redundant because it sticks to the topic redundantly. But if 'and' systematically introduces its own subject matter, we should be able to concoct such a situation. Relatedly, it is difficult to think of a discourse context where 'Jane is a lawyer' and 'John is a lawyer' would both be on-topic, but their conjunction would not be. Denying Transparency presumably commits one to such contexts. *Mutatis mutandis* for 'or' (obviously so, if this is the negation of a conjunction).

Of course, claims can intuitively be about states of affairs, in which case one might wonder: is 'Jane is not late' about **Jane not being late**, in contrast to 'Jane is late'? This is hardly obvious. Merely saying that Jane is not late is not to say anything *about* Jane not being late: 'Jane is not late' is a strange response to the proposal 'Let's talk about Jane not being late'. More natural responses: 'Jane being on-time is more important than Jean being on-time', 'Jane not being late was a pleasant surprise'.

If Transparency held for any piece of vocabulary we are willing to call 'logical', it would accommodate the venerable idea that the laws of logic are formal in the sense of being topic-neutral, or subject-matter-independent (MacFarlane 2017, sect. 4). For instance, $(\varphi \wedge \psi) \supset \varphi$ (with '$\supset$' the material conditional, defined the usual way out of negation and disjunction, or negation and conjunction) captures a logical fact, invariant whether our topic is **mathematics**, **botany** or **Jane's profession**.

However, it's uncertain whether all the vocabulary we may want to call 'logical' is topic-transparent. Surely the topic-sensitive intentional operators to be introduced in subsequent chapters are not – this issue will be briefly discussed in Section 3.1. One may want to conclude that they are not logical, and plausibly so, since they express the ascription of mental states or attitudes to agents. It may be more problematic, however, to subtract the box of necessity, at least in some alethic readings, from the logical bunch. But 'Necessarily, John is human' seems to address a different topic from 'John is human' in a number of natural conversational contexts. Some have proposed that the distinction between logical and non-logical vocabulary comes in degrees, because topic-neutrality comes in degrees: Lycan (1989) thinks that the extensional sentential connectives are more topic-neutral than the alethic modal and temporal operators, which are more topic-neutral than the epistemic modal operators. We won't discuss the issue further, except for mentioning that it quickly becomes very complex, involving deep issues on the nature of logic: again, see MacFarlane (2017) for an introduction.

Being a plausible view at least for the connectives that get the standard truth-functional treatment in classical logic, Transparency issues a ruling on other cases where intuition is murky. The hard-nosed CEO says: 'In this meeting, we will only talk about what we *should* do. Save talking about what we *needn't* do for the next'. Isn't 'Jane should phone the client' on-topic here, while 'Jane need not phone the client' is not? But the latter is equivalent to the negation of the former.

Well, it's fine for the CEO to want to prioritize the obligatory. But if the claim is literally coherent, 'Jane should phone the client' is on-topic but its presumed negation – 'Jane need not phone the client' – is not. Sensibly, Transparency disagrees: if Ann says to the CEO 'Jane should phone the client', Bob can challenge this with 'No, Jane need not phone the client' without going off-topic. In the CEO's statement, literal semantic meaning and what's pragmatically communicated differ.

2.2.3   (Co-)Necessities About Different Things

It seems that pairs of sentences can be about different things although they are both necessary (of the same sort of necessity). Here's a list of cases:

8. Water is H2O.

9. Socrates is human.

These are, for many essentialists, metaphysical necessities (in the narrow sense: see the first footnote to this chapter). Only one is about **water**.

10. $2 + 2 = 4$.

11. $3 + 3 = 6$.

These (besides plausibly being absolute necessities) are *a priori* truths if anything is: epistemic necessities requiring no empirical evidence to be eliminated. Only one is about **2**.

12. Manifolds are topological spaces.

13. Normal modal logics are closed under the rule of Necessitation.

These are unrestrictedly necessarily true, if standard definitions of this kind are. Only one is about **manifolds**.

14. If swans are white, then swans are white.

15. If $2 = 2$, then $2 = 2$.

These are logical necessities. Only one is about **swans**.

16. If someone is horribly late, then someone is late.

17. If someone is a brother, then they are a sibling.

One may call these 'analytic necessities', knowable merely in virtue of understanding their meaning. Only one is about **horribly late people**. (Analyticity has been in trouble since Quine; but if there are analytic truths at all, these would be, we think, good candidates.)

Then the following is not a valid principle for the box $\Box$ of (metaphysical, mathematical, epistemic, definitional, logical, analytic) necessity:

18. If $\Box\varphi$ and $\Box\psi$ then $\varphi \approx \psi$.

It also seems that pairs of contingent sentences can be about different things although they are true/false in the exact same circumstances ('co-necessary'):

19. Matt is a *communist*.

20. *Matt* is a communist.

One may think that these can differ in topic because they answer to different issues or questions. The former is more appropriate for 'What are Matt's political views?'; the latter, for 'Do you know any real-life communist?'.

21. Jane is a lawyer.

22. That Jane is a lawyer is true.

Only the latter can naturally be about the proposition that Jane is a lawyer.

23. Socrates exists.

24. {Socrates} exists.

The latter is about a set; the former is not.

25. Hillary Clinton is self-identical.

26. Hillary Clinton is not an abstract object.

Only the latter speaks to the issue: is Hillary Clinton abstract? (Perhaps these should count as true in all circumstances where Clinton exists, and this would make them absolutely necessary. Then they could be moved to the first group above.)

Then the following is not a valid principle for $\square$ either, where $\equiv$ is the material biconditional:

27. If $\square(\varphi \equiv \psi)$ then $\varphi \approx \psi$

2.2.4 Topics vs. Guises

Aboutness connects to hyperintensional aspects of language. So do such things as guises or modes of presentation. How do topics and guises differ? The question matters for us also because, as already anticipated in chapter 1 and developed in subsequent ones, the hyperintensionality delivered by topic-sensitivity will be used to model certain aspects of the hyperintensionality of our thoughts, such as our sometimes believing exactly one of two necessarily equivalent contents.

Now guises or modes of presentation have typically been introduced to account for the well-known opacity of attitude ascriptions, in particular due to failures of Substitutivity for (Rigid) Co-Referential Terms: see Nelson (2019) for a rich overview. When 'Cicero' and 'Tully' refer to the same person, 'Tully fell out of bed' and 'Cicero fell out of bed' are intensionally equivalent due to the fact that 'Cicero' and 'Tully' designate rigidly. But, according to many philosophers of language, they may differ in terms of the mode of presentation or guise for the famous Roman orator; and this helps with the *prima facie* difference in truth value between 'Caesar believes that Cicero fell out of bed' and 'Caesar believes that Tully fell out of bed', in situations in which Caesar is unaware of the fact that Cicero just is Tully.

To begin with, we've been taking topics as properly semantic items, in the sense that they are constituents of propositional contents. Instead, one may just not want to take guises, or modes of presentation, as aspects of meaning at all – one may take them as pragmatic devices, especially if one is a Millian or direct reference theorist on names: see, e.g., Salmon (1986)'s distinction between 'semantically encoded' and 'pragmatically imparted' information, in the context of his introduction of representational guises for the purpose of dealing with Frege's puzzle of informative identities. The debate on guises quickly gets complicated here. E.g., one may take it as an advantage of Salmon's guises that they can account for certain compositional phenomena involving embeddings. But this may make them look too similar to Fregean senses in disguise, that is, things the direct reference theorist does not want to belong in the semantics of names properly so called. A critique along these lines can be found in Forbes (1987). For a retort, see Branquinho (1990).

But even if one takes guises as properly belonging in semantics, they are typically supposed to be activated only in intentional contexts. Instead, an argument from Perry (1989) may be used to point out that topic-sensitivity shows up elsewhere, too. 'Caesar brought it about that' is no attitude ascription. It creates a hyperintensional, topic-sensitive context:

28. Caesar brought it about that Tully fell out of bed.

29. Caesar brought it about that both Tully fell out of bed and Tully is self-identical.

Presumably, 'Tully fell out of bed' and 'Tully fell out of bed and Tully is self-identical' are necessarily equivalent. However, (28) can be true while (29) is not (one cannot bring it about that an instance of self-identity is the case). It seems that the truth of 'Caesar brought it about that φ' is sensitive to the subject matter of φ.

But when 'Cicero' and 'Tully' name the same person, it cannot happen that (28) is true while (30) is not:

30. Caesar brought it about that Cicero fell out of bed.

So we seem to have a topic-sensitive, not guise-sensitive operator. There are more:

31. That Tully fell out of bed is true because Tully fell out of bed.

32. Tully fell out of bed because Tully fell out of bed.

Necessarily, that Tully fell out of bed is true just in case Tully fell out of bed. But while (31) rings true provided truth is grounded in facts, (32) is false, if facts don't explain themselves. 'Because' would, then, be sensitive to subject matter: while 'That Tully fell out of bed is true' speaks about a proposition, 'Tully fell out of bed' doesn't.

But when 'Cicero' and 'Tully' name the same person, it cannot happen that (33) is true, but (34) is not:

33. Tully feels pain because Tully fell out of bed.

34. Cicero feels pain because Tully fell out of bed.

Thus, 'because', too, seems to be topic-sensitive, not guise-sensitive.

Such considerations are not conclusive, e.g., see Dan Marshall (2021) for a recent push-back against the Perry 'bringing it about' argument. Perhaps they merely reflect some elusive difference between expressions like 'Caesar believes that' and 'Caesar brought it about that', rather than between guise and topic. Further thorny questions are afoot, e.g., is guise-sensitivity best incorporated into a theory of propositions (Chalmers 2011), or by treating intentional attitude operators as a three-place relation between an agent, proposition, and guise (Crimmins 1992)?

All this in mind, we delay further determining the relation between topic and guise for elsewhere. One reason for doing so is that this book is only concerned with *de dicto* intentionality, and puts the topicality of whole propositions (the contents of *de dicto* states), and of the sentences expressing them, at centre stage. On the other hand, even if guise-sensitivity does show up in ascriptions of *de dicto* intentional states as well ('Caesar believes that Cicero is Cicero' vs. 'Caesar believes that Cicero is Tully', etc.), and so intentional contexts may well be both guise- and topic- sensitive, it seems to be more strictly connected to the aforementioned Substitutivity of (Rigid) Co-Referential Terms, that is, of subsentential components of sentences. And, as was noted already in the introduction to chapter 1 and will be picked up again at the end of chapter 5, how topicality and subject matters should work for the subsentential components of sentences is, at present, a less developed area of aboutness research – although we mentioned there Hawke (2018) and Badura (2021b).

2.3   1C Semantics

1C, recall, is a disjunctive reducibility claim: either topics reduce to truth conditions, or vice versa. We focused on the more roundabout disjunction of:

(1C-a) Knowing C_P is *a priori* sufficient for knowing T_P (there's an f such that $T_P = f(C_P)$), for every proposition P.

(1C-b) Knowing T_P is *a priori* sufficient for knowing C_P (there's an f such that $C_P = f(T_P)$), for every proposition P.

How could 1C be wrong? Here follows a *prima facie* case for truth conditions and topics being irreducible to each other.

2.3.1   Dr X and Dr Y

It seems that one may sometimes know the truth conditions for φ without being in a position to know, absent further empirical information, what φ is about, or vice versa. Here's a twofold (a-b) situation involving epidemiologists Dr X and Dr Y, who also happen to be nerdy logic amateurs.

(a) They are observing John Jackson and Jack Johnson. Y doesn't know which of John and Jack is Jackson. X says φ: 'Either Jackson is infected, or he isn't'. Y is logically astute and knows what φ's truth conditions are: it must be true under any circumstance. But Y does not know if φ is about **John** or **Jack**.

This is no failure to think hard: to settle the matter, Y needs further empirical information on who X is talking about. Why? One simple explanation would be that there are propositions P and Q, which are live empirical candidates for what φ expresses, with P and Q having identical truth conditions but different topics.

The scenario does not assume that Y cannot grasp the content of φ. Y is positioned to know both that John is either infected or not, and that Jack is either infected or not. Y is not positioned to settle *a priori* what content attaches to φ. But whatever that content is, Y is already on top of it.

(b) Dr X and Dr Y are now observing John. X says φ: 'It is not that it is not that it is not that it is not that either John is infected or not'. Y hears this, but loses track of the number of negations before 'Either John is infected or not' (there are four). Y knows what φ's subject matter is: it's the same as that of 'Either John is infected or not'. Y is logically astute and knows that this is true under any circumstance. But Y ignores the truth conditions of φ: Y does not know if it is true in any circumstance, or in none.

This is no failure to think hard: to settle the matter, Y needs further empirical information on the form of X's utterance. Why? One simple explanation would be that there are propositions P and Q, which are live empirical candidates for what φ expresses, with P and Q having identical topic but different truth conditions.

A (1C-a) theorist should say something about (a), while a (1C-b) theorist should say something about (b). We start with the latter, for it's easier to deal with.

2.3.2   1C and Transparency

A (1C-b) theorist may reply that the missing information on the number of negations in X's utterance in (b) makes a difference with respect to its topic. This highlights how (1C-b) stands in tension with Transparency. If there's an f such that $C_P = f(T_P)$, then because Transparency mandates $\lceil \neg\varphi \rceil = \lceil \varphi \rceil$, $f(\lceil \neg\varphi \rceil) = f(\lceil \varphi \rceil)$: φ and $\neg\varphi$ are true in the same circumstances. That can't be right, so given Transparency, one would have to reject (1C-b).

Disputing just Negation Trasnsparency won't help if the connectives are Boolean, for then they are inter-definable. One would need to dispute the transparency of the Sheffer stroke, or that $\varphi \supset \varphi$ is just about what φ is about, where the horseshoe is the material conditional. (This draws on the argument Yablo (2014), sect. 2.8, gives for rejecting a specific 1C view. We generalize.)

2.3.3 1C and Topic-Diverging (Co-)Necessities

A (1C-a) theorist may reply that the missing information on what X's utterance is about in (a) makes a difference with respect to truth conditions. What the case shows, a (1C-a) theorist may say, is that we need to give truth conditions by focusing on the right kind of circumstances. (1C-a) is more work than (1C-b). We start by speaking a bit in favor of having around topic-divergent (co-)necessities.

Say the truth conditions for P are aptly modelled using sets or aggregates of circumstances: those at which P is true (and, those at which it is false, if these are to be specified separately). Whatever kind of circumstance one works with (worlds, states of affairs, situations, or whatnot), the semantics will deliver a notion of necessity: say that φ expresses a *semantic necessity* just in case it is true at unrestrictedly every circumstance. Correspondingly, say that φ and ψ express semantic co-necessities if they coincide in truth value at unrestrictedly every circumstance. If claims like 'If Jane is horribly late then Jane is late' and 'If John is a brother, then he's a sibling' count as semantic necessities, (1C-a) entails that they have identical subject matter. If claims like 'Jane is a lawyer' and 'That Jane is a lawyer is true' are semantic co-necessities, (1C-a) entails that they have identical subject matter. That seems wrong.

We saw in Subsection 2.2.3 that there seem to be various *prima facie* cases of (co-)necessary claims that diverge in what they talk about. Can one deny topic-divergent semantic (co-)necessities across the board? A denier must somewhere break with the following reasoning. Semanticists widely agree that one job of a semantic theory is to recover patterns of entailment. Ordinary judgements of entailment typically serve, as canonical works have it (Chierchia and McConnell-Ginet 1990; Heim and Kratzer 1997), as data for semantic theorizing: from 'Jane is not late' to 'It is false that Jane is late' (plausibly, only the latter is about a proposition); or from 'Barack Obama has never drunk whiskey' to 'Barack Obama has never drunk whiskey with Dolly Parton' (only the latter is about **Dolly Parton**).

Semantic theory must also recover certain entailment forms: for instance, that from $\varphi \wedge \psi$ to φ, and from φ to $\varphi \wedge \psi$ when φ entails ψ: if φ entails ψ then φ and $\varphi \wedge \psi$ are mutually entailing. But the circumstances that fix truth conditions determine the entailments relevant to semantic theorizing: φ entails ψ just in case every circumstance that renders φ true also renders ψ true (and every circumstance that renders ψ false also renders φ false). Thus, if φ entails ψ then φ and $\varphi \wedge \psi$ are semantic co-necessities. In particular, 'Jane is not late' and 'Jane is not late and it is false that she is late' are topic-divergent semantic co-necessities. Ditto for 'Obama has never drunk whiskey' and 'Obama has never drunk whiskey and he has never drunk whiskey with Parton'.

Topic-divergent semantic (co-)necessities seem to explain a lot. 'Jane is horribly late, but not late' sounds marked: it's difficult to think of a natural context where such a claim would be felicitous when taken literally. (Surely 'Jane's not late – she is *horribly* late!' can be uttered by one who wants to stress that only claiming 'Jane's late' does not quite convey the seriousness of the delay; if questioned, though, the utterer would confirm that the person *is* late.) Same for 'Jane is a lawyer but it is not true that Jane is a lawyer'.

Besides, some of the necessities in our Subsection 2.2.3 above are easily recognizable as truths across diverse contexts. They seem easily knowable *a priori*, if any claim is. Similarly, knowledge of the first in some of our pairs of co-necessities seems tantamount to knowledge of the second, for agents equipped with the relevant concepts. Competent speakers will feel puzzled when pondering the denials of our sample necessities. Nor does this merely signal entrenched knowledge: 'Obama is married' is well-entrenched, but its denial is obviously semantically perspicuous. The mere content of various of those claims apparently makes it plain to a speaker with a suitable grip on that content that they express (co-)necessities. It seems relatively easy to acquire such a grip. That's because the (co-)necessities in question are (i) semantic (co-)necessities (ii) readily recognized by competent speakers.[3]

(1C-a) theorists, however, may be able to account for all of these considerations. The idea would be, roughly: one needs to fine-grain our circumstances in the right way. Surely, if by 'circumstances' one means coarse-grained, classical, maximally consistent possible worlds, one will have a hard time telling apart the truth conditions of various necessary or co-necessary sentences that seem to differ in topic. But if we make use of the right kind of circumstances, everything will fall into place: we will have an account of truth conditions that distinguishes the pairs in Subsection 2.2.3, in a way that offers both a satisfying account of pre-theoretic natural language entailment, and a satisfying account of the special status of our alleged semantic necessities. This is to account for part (a) of the Dr X and Dr Y case too: what Dr X's claim is about makes a difference with respect to the circumstances under which it is true/false – once 'circumstance' has been properly fine-tuned.

What should we take as our working circumstances, then? Suppose one takes an especially liberal attitude: for any two syntactically distinct sentences φ and ψ, there exists a

---

[3]This echoes the traditional starting point of philosophers' vexed discussion of 'analytic truth' (Rey 2018): certain ordinary claims strike us, pretheoretically, as (i) expressing necessities, (ii) easily knowable, and (iii) governed by meaning facts that make (i) and (ii) plain. We agree with the tradition that there is something to explain here. This is a modest commitment. We are not thereby committed to the existence of 'analytic truths', traditionally understood, i.e. sentences with non-linguistic content that are *made* true by mere linguistic facts: cf. Quine (1976); Boghossian (1996); Russell (2008). Nor to the claim that understanding a semantic necessity entails that one knows it: cf. Boghossian (1996); Williamson (2007). Nor that having ordinary linguistic competence with respect to φ entails that one knows its truth conditions and topic. Nor that negated semantic necessities cannot be believed. Nor that negating any two semantic necessities yields the same content: cf. Stalnaker (1984); Field (1986b).

circumstance in which either is true and the other isn't. This can be obtained by using 'open' impossible worlds, i.e., worlds not closed under any non-trivial notion of entailment, as our circumstances (Priest 2016; Berto and Jago 2019). 'Jane is a lawyer and an expert footballer' holds at some point where 'Jane is a lawyer' doesn't. 'Jane is horribly late' holds at some point where 'Jane is late' doesn't.

If the open worlds approach were proposed as a way of giving a general account of propositional content, it would trivialize the appeal to circumstances: see Berto and Jago (2019), 114, for a discussion of the issue in the context of epistemic logics indiscriminately using open impossible worlds semantics. The only constraint on a circumstance that renders φ true is that φ is stipulated as holding true there. It would push the idea of using circumstances to model truth conditions to the breaking point: that 'Jane is a lawyer and an expert footballer' is true only if 'Jane is a lawyer' is true reflects an obvious fact about the truth conditions of conjoined sentences. And the account would treat 'Jane is horribly late but not late' and 'Obama is not married' symmetrically: one who grasps their respective content should judge them equally semantically felicitous, for the truth conditions for each admit circumstances where it's true and circumstances where it's not.

Suppose one takes circumstances as partial states or situations (Barwise and Perry 1983). The truth conditions for φ are composed of those states that necessitate φ's truth and those that necessitate its falsity. One may then (i) avoid positing semantic necessities while (ii) explaining our natural attitude to various necessities in our list. For (i): one denies that 'If Jane is horribly late then Jane is late' and 'Manifolds are topological spaces' are made true by every partial situation (e.g., those that don't include Jane, or manifolds). For (ii): one claims that neither has any *false*makers, and that our easily appreciating this explains why we judge various among them necessary, or easily knowable, etc.

What is necessitation? Here's one account that won't work: state s necessitates φ just in case φ is true at every possible world with s as a part. It follows that any absolute or metaphysical (in the sense) necessity φ is made true by every state, so it's a semantic necessity. '$2 + 2 = 4$' is made true by arbitrary state s since '$2 + 2 = 4$' is true at every possible world that embeds s.

One may retreat by admitting impossible worlds: s necessitates φ just in case φ is true at every world, possible or impossible, with s as a part. Then '$2 + 2 = 4$' is not a semantic necessity: there is an impossible world where $2 + 2$ is not equal to 4. 'Socrates exists' and '{Socrates} exists' are not semantically co-necessary: there is an impossible world where Socrates exists but the singleton does not. This would be a slippery slope to open worlds liberalism: once certain impossible worlds are admitted, why not all?

Anyway, applying the strategy to our full range of examples contradicts plain truth-conditional facts: our hypothetical state theorist must allow worlds where 'Manifolds are topological spaces' is false; where 'John is a married bachelor' is true; where 'Jane is a lawyer' is true but 'That Jane is a lawyer is true' is false, or vice versa. If this retreat is chosen, the method of representing mere truth conditions with a state space is abused: obvious constraints imposed by the meanings of 'manifold', 'bachelor', and 'It is true that', are ignored. For related reasons, truthmaker theorists roundly deny that the aforementioned reading of 'necessitation' captures the most interesting sense in which states necessitate truth (Armstrong 2004, sect. 2.3).

A way better alternative understands truthmaking in terms of (fundamental) metaphysical explanation: Jago (2018), ch. 6; see also Schipper (2018, 2020) for cognate ideas. And connects it to a certain notion of exactness. A truthmaker for φ is *exact* when 'it can necessitate the sentence while being wholly relevant to its truth' (Fine and Jago 2018, sect. 1). A truthmaker for φ is *inexact* when it contains an exact truthmaker for φ as a part. A situation in which (just) Socrates is both a philosopher and married is an exact truthmaker for 'Socrates is a philosopher and married', not for 'Socrates is married', since Socrates' being a philosopher is irrelevant to the truth of 'Socrates is married'. The truth conditions of φ are best modelled as a pair: its set of exact truthmakers and its set of exact falsemakers (Fine 2016a,b, 2017; Fine and Jago 2018).

With this setting in place, one can then deny that, say, '2 is a number' and 'Vixens are female foxes' are semantic (co-)necessities, because they have different exact truthmakers (e.g. the one for the former sentence has 2 as a constituent, the one for the latter doesn't). And one can still posit the semantic feature that explains our reactions to many such sentences, as sketched above: neither has a falsemaker.

Are these still truth conditions? It is one thing to theorize about the states that, if they obtained, would (exactly) explain the truth of φ; to theorize about φ's truth conditions is another. But one needn't be draconian in one's reading of 'truth conditions'. Let's say a *strict* account of the truth conditions of φ comes via a set of worlds at which φ is true. A *liberal* account of the truth conditions of φ comes via a set of states that determine the set of worlds at which φ is true. So the truthmaker theorist's account is liberal. A (1C-a) theorist of this sort can claim that for every interpreted φ there exists a set of states from which both the worlds at which φ is true and φ's topic can, if one likes, be extracted.

The proposal as such doesn't eradicate all topic-divergent semantic (co-)necessities. Say φ and ψ are topic-divergent and without falsemakers. φ: '2 is a number'; ψ: 'Vixens are female foxes'. Take $\varphi \vee \neg\psi$. Say our theorist accepts a Finean account of the exact truthmakers and falsemakers for logically complex claims (Fine 2016a): a state s is an exact falsemaker for $\varphi \vee \neg\psi$ just in case there exist states s_1 and s_2 such that s_1 is an exact falsemaker for φ, s_2 is an exact falsemaker for $\neg\psi$ and s is the fusion of s_1 and s_2. No state s satisfies this, since no s_1 serves as an exact falsemaker for φ. Thus, $\varphi \vee \neg\psi$ has no falsemakers. s is an exact truthmaker for $\varphi \vee \neg\psi$ just in case it is an exact truthmaker for either φ or $\neg\psi$ (or a fusion of one for φ and one for $\neg\psi$). The exact truthmakers for $\neg\psi$ are exactly the exact falsemakers for ψ. Thus, $\neg\psi$ has no truthmakers, and so the exact truthmakers for $\varphi \vee \neg\psi$ are just those for φ. Wrapping up: φ and $\varphi \vee \neg\psi$ have the same exact truthmakers and falsemakers. So they have the same truth conditions. But, against the 1C reduction, they have different topics: only one is about **vixens**. They are topic-divergent semantic co-necessities.

A truthmaker semanticist, however, has a way out again.

The semantics can work in such a way that *every* sentence has a falsemaker (Fine 2016b, sect. 4). Then topic-divergent semantic co-necessities cannot be constructed as in the last paragraph. One may then propose that our typical reaction to various pairs of alleged topic-divergent semantic co-necessities be explained by the absence of *possible* states that make one true but the other false.

The truthmaker theorist, then, may need absolutely impossible states of all sorts, in particular, ones that serve as falsifiers for '2 is a number', 'Vixens are female foxes', etc. (and, that serve as verifiers for 'John is a married bachelor', etc.). One may say that admitting impossible states with married bachelors, male vixens, etc., is a cost, if the account is to be truth-conditionally plausible. A basic truth-conditional fact: if 'Jane is an expert lawyer' is true then 'Jane is a lawyer' is true. Another: if 'Jane is a lawyer' is true then 'Jane is not a lawyer' is not. The present account of truth conditions, however, would allow states at which both 'Jane is an expert lawyer' and 'Jane is not a lawyer' is true, namely that include, as a part, both an exact truthmaker for 'Jane is an expert lawyer' and an exact falsemaker for 'Jane is a lawyer'. Presumably, such states are impossible precisely in virtue of violating truth-conditional facts. But if such states *mis*represent truth conditions, they should be excluded from those that accurately represent truth conditions. At this point, we (*qua* 2C theorists) would be left with a bit of an incredulous stare: it is not clear to us that this kind of 1C theorist has managed to provide plausible (liberal) truth conditions only by giving us exact truthmaking and falsemaking conditions.

We close the Section by considering a more abstract proposal for encoding topicality within truth-conferring circumstances, following Beall (2016). In a Weak Kleene framework, valuation functions assign sentences one of three semantic values (0, 0.5, 1), with the following constraints: if φ contains an embedded sentence that is assigned 0.5, then φ is assigned 0.5; else, φ is evaluated classically. For instance, $p \lor q$ and $p \land q$ are respectively assigned 1 and 0 if p is assigned 1 and q is assigned 0; $p \lor q$ and $p \land q$ are both assigned 0.5 if p is assigned 1 and q is assigned 0.5. Beall proposes we interpret 0.5 as marking sentences that are *off-topic* relative to a(n implicit) background topic; 1 as marking sentences that are on-topic and true; 0 as marking sentences that are on-topic and false. Hence, such valuation functions serve as topic-sensitive circumstances of evaluation, abstractly understood. A (1C-a) theorist might propose: we have here a truth-conditional framework from which subject matter can be recovered: the topic of φ may be identified with the set of all circumstances (i.e. valuations) for which φ receives 1 or 0. Two sentences have identical subject matter when their sets coincide; the subject matter of one contains that of the other when we have set inclusion.

However, the space of Weak Kleene valuation functions fails to plausibly capture (mere) truth conditions. Consider the entailment relation generated by the proposed logical space: a circumstance c may assign 1 to p without assigning 1 to $\neg(\neg p \wedge q)$, since q may be off-topic relative to c. This violates a (presumably) fundamental truth-conditional fact: p entails $\neg(\neg p \wedge q)$. Nor are the points in this logical space intended to represent mere truth valuations. Perhaps they implicitly represent a world or situation combined with a relevant topic.

This applies equally to more flagrant attempts to encode topicality into points of evaluation. Take a semantics where a point of evaluation is a pair: a world w and a set of proposition letters t, capturing the relevant topic for evaluation. φ receives 1 at $\langle w, t \rangle$ when φ is true at w and on-topic with respect to t (i.e. every atomic formula in φ appears in t). This has the abstract form of a standard truth-conditional framework. However, it fails to plausibly capture mere truth conditions although φ's truth conditions and its topic can be recovered from the set of points of evaluation that assign 1 to φ. The points of evaluation aren't circumstances in the intuitive sense relevant to truth-conditional semantics: they are not mere worlds, or states, or situations, etc. We seem to have a 2C framework in disguise.

2.4   2C Semantics

Here's a simple formal semantics for our language $\mathcal{L}$ above, which captures, we think, all of the constraints recommended by our topicology from Section 2.2. A *frame* for $\mathcal{L}$ is a tuple $\mathfrak{F} = \langle W, D, \mathcal{T}, \oplus \rangle$, understood as follows:

- W is a non-empty set of possible worlds.

- D is a non-empty domain of objects, which for simplicity we take as world-invariant.

- $\mathcal{T}$ is a non-empty set of topics: the subject matters the formulas of our language $\mathcal{L}$ can be about.

- $\oplus : \mathcal{T} \times \mathcal{T} \to \mathcal{T}$ is *topic fusion*: a binary operation making topics part of larger topics and satisfying, for all $x, y, z \in \mathcal{T}$:

  (Idempotence) $x \oplus x = x$

  (Commutativity) $x \oplus y = y \oplus x$

  (Associativity) $(x \oplus y) \oplus z = x \oplus (y \oplus z)$

Fusion is unrestricted: $\forall xy \in \mathcal{T} \; \exists z \in \mathcal{T}(z = x \oplus y)$. One can then define topic parthood or topic inclusion, $\leq$, from $\oplus$ as per above: $x \leq y := x \oplus y = y$. Thus, it's a partial ordering – for all $x, y, z \in \mathcal{T}$:

  (Reflexivity) $x \leq x$

  (Antisymmetry) $x \leq y \;\&\; y \leq x \Rightarrow x = y$

  (Transitivity) $x \leq y \;\&\; y \leq z \Rightarrow x \leq z$

Then $\langle \mathcal{T}, \oplus \rangle$ is a join semilattice, and $\langle \mathcal{T}, \leq \rangle$ a poset.

A *model* $\mathfrak{M} = \langle W, D, \mathcal{T}, \oplus, \mathsf{d}, \mathsf{t} \rangle$ adds to a frame two semantic functions. The first one, d, takes care of truth conditions: it assigns a denotation to each term, predicate, and sentence of $\mathcal{L}$. For a constant a, $\mathsf{d}(a) = \mathsf{a} \in D$. For a predicate symbol R, $\mathsf{d}(R) = \mathsf{R}$ is an intension, taking a world and returning a set of pairs of objects in D. For a sentence φ, $\mathsf{d}(\varphi) = \lfloor \varphi \rfloor \subseteq W$, with the (standard) truth-functional constraints specified below.

The second function, t, takes care of topics: it assigns a topic from $\mathcal{T}$, or a recipe for generating a topic from given topics, to each term, predicate, and sentence in $\mathcal{L}$. For a constant a, $t(a) = \mathbf{a} \in \mathcal{T}$: intuitively, 'Jane' is assigned the topic **Jane**. For a predicate symbol R, $t(R) = \mathbb{R}$ is a function taking a pair of topics and returning a topic in $\mathcal{T}$: for names 'Jane' and 'Joan', $\text{kicked}(\mathbf{Jane}, \mathbf{Joan})$ returns the topic expressed by the question, 'Did Jane kick Joan?'. For a sentence φ, $t(\varphi) = \lceil \varphi \rceil \in \mathcal{T}$. This, as well as the truth-functional interpretation, has to satisfy the following compositional constraints:

- $\lfloor aRb \rfloor = \{ w \in W | \langle \mathbf{a}, \mathbf{b} \rangle \in \mathbb{R}(w) \}$

- $\lceil aRb \rceil \leq \mathbb{R}(\mathbf{a}, \mathbf{b}) \leq t(a) \oplus t(b)$

- $\lfloor \neg \varphi \rfloor = W \setminus \lfloor \varphi \rfloor$

- $\lceil \neg \varphi \rceil = \lceil \varphi \rceil$

- $\lfloor \varphi \wedge \psi \rfloor = \lfloor \varphi \rfloor \cap \lfloor \psi \rfloor$

- $\lceil \varphi \wedge \psi \rceil = \lceil \varphi \vee \psi \rceil = \lceil \varphi \rceil \oplus \lceil \psi \rceil$

- $\lfloor \varphi \vee \psi \rfloor = \lfloor \varphi \rfloor \cup \lfloor \psi \rfloor$

Next, $\trianglerighteq$, $\Box$, $\approx$ are treated as global operators:

- $\lfloor \varphi \trianglerighteq \psi \rfloor = W$, if $\lfloor \varphi \rfloor \subseteq \lfloor \psi \rfloor$ and $\lceil \psi \rceil \leq \lceil \varphi \rceil$. Else: $\lfloor \varphi \trianglerighteq \psi \rfloor = \emptyset$.

- $\lfloor \Box \varphi \rfloor = W$, if $\lfloor \varphi \rfloor = W$. Else: $\lfloor \Box \varphi \rfloor = \emptyset$.

- $\lfloor \varphi \approx \psi \rfloor = W$, if $\lceil \psi \rceil = \lceil \varphi \rceil$. Else: $\lfloor \varphi \approx \psi \rfloor = \emptyset$.

We only give truth conditions for these: how topicality should work for sentences that include operators of this kind, is an interesting question better left for elsewhere. What we care about here is only the conditions under which the relevant claims are true.

Relative to $\mathfrak{M}$, a content is a pair $\langle C, x \rangle$, where $C \subseteq W$ and $x \in \mathcal{T}$. Then the content of φ is $[\varphi] = \langle \lfloor \varphi \rfloor, \lceil \varphi \rceil \rangle$. Entailment is, completely standardly, truth preservation at all worlds of all models: $\varphi_1, \ldots, \varphi_n \vDash \varphi$ if for every model $\mathfrak{M}$ and $w \in W$, if $w \in \lfloor \varphi_i \rfloor$ for every i, then $w \in \lfloor \varphi \rfloor$. Validity is, equally standardly, truth at all worlds of all models: $\vDash \varphi$ if for every model $\mathfrak{M}$ and $w \in W$, $w \in \lfloor \varphi \rfloor$.

We then get some notable validities and invalidities (the proofs are easy):

35. $\vDash (\varphi \wedge \psi) \unrhd \varphi$

36. $\nvDash \varphi \unrhd (\varphi \vee \psi)$

37. $\nvDash \varphi \unrhd \neg(\neg\varphi \wedge \psi)$

38. $\vDash (\varphi \wedge \psi) \unlhd\unrhd (\psi \wedge \varphi)$

39. $\vDash \varphi \unlhd\unrhd \neg\neg\varphi$

40. $\nvDash aRb \approx bRa$

41. $\vDash \varphi \approx \neg\varphi$

42. $\vDash (\varphi \wedge \psi) \approx (\varphi \vee \psi)$

43. $\Box\varphi \wedge \Box\psi \nvDash \varphi \approx \psi$

44. $\Box(\varphi \equiv \psi) \nvDash \varphi \approx \psi$

Some comments: (35) and (36) mark an essential difference between conjunction and disjunction. They tell us that, whereas the content of a conjunction includes, and not just entails, that of its conjuncts, the content of a disjunct does not perforce include that of the disjunction, in spite of entailing it: the other disjunct can bring in extra topic. Following Yablo (2014) again: $\varphi \vee \psi$ can *say less about more* than φ: the disjunction can address a larger topic than that of one of its disjuncts, even while being less informative in that it rules out fewer circumstances. Instead, $\varphi \wedge \psi$ can *say more about more* with respect to φ: it can both address a larger topic and be more informative, if it rules out more circumstances.

(35) and (36) are a widely recognized mark of a topic-sensitive semantics:

> A paradigm of inclusion, I take it, is the relation that simple conjunctions bear to their conjuncts – the relation *Snow is white and expensive* bears, for example, to *Snow is white*. A paradigm of noninclusion is the relation disjuncts bear to disjunctions; *Snow is white* does not have *Snow is white or expensive* as a part. (Yablo 2014, 11)

> A guiding principle behind the understanding of partial content is that the content of A and B should each be part

of the content of $A \land B$ but that the content of $A \lor B$ should not in general be part of the content of either A or B. (Fine 2016a, 200)

(35) and (36) will be very important in subsequent chapters: they will deliver (arguably) nice logical closure and non-closure features for our TSIMs. Failures of Addition, (36) will be discussed at length. In such contexts, we shall talk about (37) as well, but since we mentioned this in connection to our discussion of Beall (2016) above, we should already remark the following: the entailment from φ to $\neg(\neg\varphi \land \psi)$ holds in our framework. What (37) says, is that this is not content-preserving: entailment does not encode topic-preservation.

(38) gives the welcome result that 'John is tall and handsome' and 'John is handsome and tall' say the same, which will also matter when intentional operators come in (try to think that John is tall and handsome without thinking that he is handsome and tall).

(39) follows from Negation Transparency. The view is shared by 2C as well as 1C approaches: see Fine (2016a) again. But we should flag that, if one goes 2C, one has reasons for not disliking part of what (39) says (the topicality part!), even if one is a non-classical logician who wants Double Negation Elimination (or, perhaps, even Double Negation Introduction) to fail in one's favourite logic. One may retain subject matter equivalence between what one sentence and what its double negation are about; change the truth-conditional bit in our semantics into a setting less classical than the one we have above; and thus make the twofold entailment fail in either or both directions. More generally (we owe this remark to Chris Badura), one can in principle plug in different characterizations of entailment and see what the resulting topic-sensitive semantics looks like.

The failure of what in Section 2.2 was called Constituent Equivalence, declared by (40), shows that the subject matter of (atomic) sentences is not fixed only by looking at their subsentential constituents.

(41) and (42) express Transparency; (43) and (44) guarantee the possibility of topic-diverging necessities and co-necessities. It's the smooth holding of (41)-(44) all together that gives, we think, the core of a 2C view.

2.5   Full Yablo?

Our final issue for this chapter is: should we accept Full Yablo or rather limit ourselves to Weak Yablo? Full Yablo has it that, for $\varphi \trianglerighteq \psi$ to hold, it is both necessary and sufficient that (1) φ entails ψ and (2) ψ is about topic x only if φ is about x. To reject sufficiency is to say that there's more to propositional content than truth conditions and topics. We would not be extremely worried if we had to go down that road: first, nothing in the critical discussion of 1C approaches presented above relies on accepting Full Yablo rather than Weak Yablo. Second, one does not need to endorse Full Yablo to find the topic-sensitive logics of thought presented in subsequent chapters interesting and useful in various ways: they will hopefully remain so even if there is more to content than truth conditions and topics. This said, we discuss two issues with sufficiency, which cause problems especially when paired with what we called 'immanent closure' since chapter 1, following Yablo (2014, 2017).

2.5.1   Mathematical and Logical Ignorance

One may worry that Full Yablo faces a problem of mathematical omniscience, if we accept immanent closure for knowledge, that is, the idea that knowledge is closed under content parthood: if $\varphi \trianglerighteq \psi$ then knowing that φ entails knowing that ψ.

It seems possible to know the axioms of ZF set theory without knowing, say, Cantor's theorem. But, it might be thought, don't the axioms encapsulate all of set theory's subject matter? Then they contain the subject matter of Cantor's theorem. Also, they entail it. Then Cantor's theorem is a part of the content (of the conjunction) of the axioms. Isn't this the core of venerable, e.g., neo-positivistic or, more generally, anti-Kantian views about the analyticity of mathematics? If knowledge is closed under content parts, then one who knows the ZF axioms knows Cantor's theorem. This seems wrong.

One may question the assumption that the subject matter of Cantor's theorem is wholly contained in that of the ZF axioms. Cantor's theorem is about the cardinality of an arbitrary set's power set. It speaks to the issue: Is the power set of an arbitrary set strictly larger in size than that set? It is not obvious that any (conjunction of) ZF axiom(s) is about this topic or addresses this issue. They rather entail something, Cantor's theorem, which is about this topic and in answer to this question.

To settle whether thick propositional content with immanent closure generates mathematical omniscience, we need to settle on a characterization of subject matter (perhaps in particular of the subject matter of quantified statements, since most mathematical axioms take this form). A toy example: is the subject matter of '16 + 16 = 32' included in that of '16+32 = 48'? Then Full Yablo with immanent closure for knowledge delivers a bad result: knowing the former is part of knowing the latter.

But only some theories of subject matter entail the inclusion. For instance, if the subject matter of '16 + 32 = 48' is taken to be $\{16, +, 32, =, 48\}$, the subject matter of '16 + 16 = 32' to be $\{16, +, =, 32\}$, and inclusion to be the subset relation, then the problematic inclusion follows. But it need not be delivered, for instance, by the truthmaker-based approach of Fine (2016a): the (fusions of the) exact truthmakers and falsemakers for '16+16 = 32' and '16+32 = 48' are plausibly distinct.

One may also worry that Full Yablo plus immanent closure for knowledge faces a problem of logical omniscience. Suppose a logic student knows that $\varphi \wedge \psi$ is true but hasn't grown accustomed to the material conditional: they deny that $\neg\varphi \supset \neg\psi$ is true. If the relevant connectives are transparent, as we claim, it follows that the content of $\neg\varphi \supset \neg\psi$ is contained in that of $\varphi \wedge \psi$ (we have both topic-containment and entailment). So, we must rule, erroneously, that the student knows that $\neg\varphi \supset \neg\psi$ is true. Or so the objection goes.

The last claim in this reasoning is ambiguous, however, and false on a key reading. One can read '$\neg\varphi \supset \neg\psi$ is true' as a metalinguistic claim: a certain sentence in a certain formal language is claimed to be true. Certainly, by the lights of Full Yablo plus immanent closure for knowledge, if our student knows the content of sentence $\varphi \wedge \psi$, then they know the content of sentence $\neg\varphi \supset \neg\psi$: the latter is part of the former. But one could say that they do not know that the sentence $\neg\varphi \supset \neg\psi$ expresses the latter content. So, in an important sense, the student can fail to know that $\neg\varphi \supset \neg\psi$ is true, as desired. Courses in basic propositional logic disseminate some metalinguistic knowledge: although the proposition that Jane is a lawyer or Jane is not a lawyer is about **Jane**, and not about everything, to learn that 'Jane is a lawyer or Jane is not a lawyer' is a tautology is to learn something about the behaviour of certain connectives, rather than about **Jane**.

This is, of course, a version of Stalnaker's metalinguistic approach to evading problems of logical omniscience (Stalnaker 1984). His strategy has received abundant criticism (Field 1978, 1986a,b; Speaks 2006). One of us (FB) has argued elsewhere (Berto and Jago 2019, sect. 8.2), that the Stalnakerian approach cannot work in general: it is implausible that, whenever one seemingly fails to know a logical or necessary truth, one is confused about what the sentence in question means.

Such an application of the strategy would be much milder than Stalnaker's, however. A key criticism, for instance, is that he reduces the accrual of mathematical knowledge to the accrual of metalinguistic knowledge. The considerations on mathematical omniscience presented above, however, show that, depending on the chosen account of subject matter, one may not need to apply the metalinguistic strategy to purely mathematical knowledge.

2.5.2   Classicality

Here's a simple and deep difficulty we have been made aware of by Alexandru Baltag. Our formal 2C semantics from Section 2.4 yields these two entailments:

45. $\Box\varphi \vDash (\varphi \vee \neg\varphi) \unrhd \varphi$

46. $\Box(\varphi \supset \psi) \vDash (\varphi \wedge (\psi \vee \neg\psi)) \unrhd \psi$

Less formally:

47. If φ is necessary, then to say $\varphi \vee \neg\varphi$ is to partly say φ.

48. When φ necessarily implies ψ, to say $\varphi \wedge (\psi \vee \neg\psi)$ is to partly say ψ.

These follow from Full Yablo and independently amenable principles. Take (47). Say φ is necessary: true at every circumstance. Assume that $\varphi \vee \neg\varphi$ is, too. Further, assume Transparency. Then φ and $\varphi \vee \neg\varphi$ have the same subject matter. By the right-to-left direction of Full Yablo, the content of φ is included in that of $\varphi \vee \neg\varphi$.

Take (48). Suppose φ entails ψ. So $\varphi \supset \psi$ is necessary: no φ-circumstance fails to be a ψ-circumstance. It follows that $(\varphi \wedge (\psi \vee \neg\psi)) \supset \psi$ is necessary as well. By Transparency, the topic of $\varphi \wedge (\psi \vee \neg\psi)$ includes that of ψ. Hence, by the right-to-left direction of Full Yablo, the content of ψ is included in that of $\varphi \wedge (\psi \vee \neg\psi)$.

On the face of it, (47) and (48) are problematic. Cantor's theorem (CT) is, plausibly, necessary. By (47), to say that either CT holds or it doesn't is to partly say that CT holds. Suppose that CT is entailed by (the conjunction of) the ZF axioms. By (48), to say that both the ZF axioms hold and either CT holds or it doesn't is to partly say that CT holds. If we accept immanent closure for knowledge, it seemingly follows from Full Yablo that knowing that either CT holds or it doesn't entails knowing that CT holds. Likewise, it seemingly follows that if one knows both the ZF axioms and that either CT holds or it doesn't then one knows that CT holds. Knowledge of CT cannot be acquired so easily.

Biting the bullet and accepting (47) and (48) is more palatable if one, finally, lets immanent closure for knowledge go. One claims that (47) and (48) involve non-obvious facts about what is said. Perhaps, although saying that either CT holds or it doesn't *is* just to say CT, we don't ordinarily realize this. But, in later chapters *we* have a lot of use for the idea of immanent closure in a general setting – for it allows us to deliver a number of interesting results in the logic of our TSIMs.

One route to accepting (47) and (48), while keeping immanent closure for knowledge, may try to deflate their import by denying that substantive truths like CT count as necessities in the relevant sense. Obvious 'semantic' necessities and entailments are epistemically insubstantial. Someone with the requisite semantic competence finds it easy to know that all bachelors are male. Likewise for: 'Either all bachelors are male or some are not'. So it isn't particularly jarring to claim that one knows (says) all bachelors are males just in case one knows (says) either all bachelors are male or some aren't.

If mathematical truths are absolutely necessary, one would, then, accept an essential role for (broadly) metaphysically or absolutely impossible circumstances. To deny that CT is such an obvious semantic necessity is to admit a circumstance at which CT doesn't hold in the space of circumstances appropriate for capturing truth-conditional facts. This circumstance would be absolutely impossible. 2C theorists can live with this: they do not thereby admit more problematic instances of impossible circumstances, e.g., impossible circumstances where entailment laws are violated, and there are independent arguments for admitting *some* absolutely impossible circumstances into one's semantic machinery (Fine 2016b).

But even granting the advocate of Full Yablo this strategy for (47), it does not transfer convincingly to (48). If one denies that 'ZF axioms ⊃ CT' expresses a necessity, one must admit circumstances where the ZF axioms hold but CT doesn't. Given that CT follows from the ZF axioms by pure logic, these would be logically impossible circumstances. We are sliding toward unrestricted logically impossible worlds liberalism. (48) shows, we think, more clearly than any other difficulty discussed in this final Section, the cost of accepting Full Yablo. (We have further discussion in Hawke et al. (2020).)

As should be clear, such a tangle of problems partly stems from the fact that we have been sticking with classical, absolutely, and, in particular, logically possible worlds in our own account – hence the label 'Objection from Classicality': taking these as our working circumstances delivers that CT is necessary, i.e., it holds at all of them; and so does the implication from the ZF axioms to CT. We will come back to this in chapter 6, particularly Section 6.3, where non-classical logical settings admitting logically impossible worlds will get some revenge over the classical possible worlds semantics we'll have employed throughout for our TSIMs. To get there, we need to start introducing our TSIMs in a proper way. This we do in the next chapter.

2.6   Chapter Summary

When do two sentences say the same thing, that is, express the same propositional content? This chapter has proposed two-component (2C) semantics, the view that propositional contents comprise pairs of irreducibly distinct components: (1) truth conditions, and (2) topic or subject matter. We have presented an abstract 2C formal semantics, which gives same-saying conditions while being neutral on the exact nature of subject matter. We have contrasted 2C with 1C semantics, the view that either truth conditions are reducible to subject matter or vice versa. In order to do so, we have developed a 'topicology', laying out a series of constraints any good theory of subject matter should obey. Finally, we have explored some difficulties for the idea that coincidence in truth conditions and topic is both necessary and sufficient for content identity, especially in connection with the idea of immanent closure.

3

Topic-Sensitive Intentional Modals

A *Topic-Sensitive Intentional Modal* (TSIM) is a modal operator representing an attitude ascription, and whose semantics features an aboutness or topicality constraint. In this chapter, and in the three following it, we will focus on two-place TSIMs of the form '$X^\varphi\psi$'; a generic reading may be something like: 'Given φ, the agent Xs (or would X) that ψ', X being some attitude. Such two-place operators can do nice things for mainstream and formal epistemology, belief revision theory, and mental simulation theory. In these four chapters, when I talk about TSIMs in general, I will always refer to two-place operators of this kind. I start with a basic formal semantics for them, and discuss some notable validities and invalidities it delivers.

The basic semantics is a TSIM analogue of that for the basic system K of normal modal logic. As is well known, in such a system endowed with the standard Kripkean possible worlds semantics, one characterizes the box of necessity as a restricted quantifier over possible worlds: $\Box\varphi$ is true at world w iff φ is true at all worlds w_1, such that wRw_1, that is, such that they are accessible from w via the binary accessibility relation R: see, e.g., Priest (2008), ch. 2. These worlds represent alternative possibilities from the viewpoint of w. Such a setting already validates a number of inferences involving the box, e.g., $\Box(\varphi \wedge \psi)$ entails $\Box\varphi \wedge \Box\psi$, with no

Topics of Thought: The Logic of Knowledge, Belief, Imagination.
Francesco Berto, Oxford University Press. © Francesco Berto 2022.
DOI: 10.1093/oso/9780192857491.003.0003

special constraint on R. Next, one gets stronger modal logics by adding conditions on R, which also make plausible different interpretations of the box itself: see, e.g., Priest (2008), ch. 3. For instance, by adding the condition that R be reflexive, one gets the logic T, whose characteristic validity is the T principle: $\Box\varphi \supset \varphi$ (with '$\supset$' the material conditional). This is needed if we want the box to represent some factive modality. Moving into the territory of Hintikka-style modal-epistemic logics, the box symbol is often replaced by a K when one reads '$K\varphi$' as saying that the agent knows that φ. We then want the T principle to hold, for knowledge implies truth. But we don't want it to hold if we are to read the box symbol, which is then often replaced by a B, as expressing belief: it should not be a validity that $B\varphi \supset \varphi$, for one can believe falsities.

Similarly, in the three chapters following this one we will explore three readings of '$X^\varphi\psi$' one gets by imposing different constraints on the accessibility relations (or, as we will see, functions) that show up in the truth conditions for the TSIMs:

(i) In chapter 4, we investigate Knowability Relative to Information (KRI), inspired by Dretske's view that what one can know depends on the available (empirical) information; we then replace the X with a K, and read '$K^\varphi\psi$' as: 'Given total (empirical) information φ, one would be in a position to know ψ'.

(ii) In chapter 5, we investigate Reality-Oriented Mental Simulation (ROMS), capturing features of mainstream mental simulation accounts; we then replace the X with an I, and read '$I^\varphi\psi$' as: 'In mental simulation starting with suppositional input φ, one imagines that ψ'.

(iii) In chapter 6, we investigate hyperintensional conditional belief or static belief revision (with hints at a dynamic development), reducing certain idealizations of cognitive agents one finds in standard doxastic logics and AGM belief revision theory; we then replace the X with a B, and read '$B^\varphi\psi$' as: 'One believes ψ, conditional on φ', or as 'If one were to learn that φ, one would believe that ψ was the case'.

3.1   The Basic Semantics

As rehearsed above, in the framework started by Hintikka (1962) ascriptions of knowledge or belief are represented via quantifiers over possible worlds, restricted by some accessibility relation R. The theory to be developed now differs from the Hintikkan framework in three main ways:

(1) The Hintikkan operators are one-place. The TSIMs in chapters 3 to 6 are two-place, '$X^\varphi\psi$'. As we will see, some one-place modals can be recovered from them by taking a triviality for φ although, in a topic-sensitive setting, that's not *quite* the same as taking a logical truth for it. (I'll get back to this in Section 6.4, and we will also see TSIMs for belief which are directly one-place in chapter 7.)

(2) The $X^\varphi\psi$s are *variably strict* modals. As a first approximation and before I make things formally precise, this means that what worlds are accessible in the evaluation of a formula of the form '$X^\varphi\psi$' at world w depends on φ: instead of having a single accessibility relation as in the Hintikkan framework, we have a bunch of them, indexed to the φ in first position in the relevant TSIM. Roughly: variability represents how different φs determine different possibilities the agent looks at (more precise readings will pop up, as we delve into specific kinds of TSIM). Variable strictness makes the operators non-monotonic in the following sense: it may happen that $X^\varphi\psi$ holds at some world w whereas $X^{\varphi\wedge\chi}\psi$ fails there, because the worlds one looks at from w given φ differ from the worlds one looks at from there given $\varphi\wedge\chi$: adding information in first position can turn a true TSIM into a false one. If you are a bit familiar with the mainstream Lewis-Stalnaker semantics for conditional logics – see Stalnaker (1968); Lewis (1973); Nute (1984) and (Priest 2008, Ch. 5) for overviews – you will have spotted that this setting makes the two-place TSIMs similar to variably strict conditionals: epistemic logic, in TSIM clothing, becomes a kind of conditional logic.

(3) The $X^\varphi\psi$s encompass a topicality or aboutness filter capturing what the relevant thoughts are (and, equally importantly, are *not*) about. As a first approximation and before I make things formally precise: what the filter requires is that ψ be fully on-topic with respect to φ, that is, the topic assigned to ψ must be fully included in that assigned to φ. The filter will have a crucial role in invalidating a number of inferences, which are valid for variably strict conditionals in standard conditional logics.

Ideas (1) and (2) are in the literature: two-place epistemic or doxastic operators expressing conditional belief, or static and dynamic belief revision ('$B^\varphi\psi$': 'Conditional on φ, one believes ψ'; '$[*\varphi]\psi$': 'After revising one's beliefs by φ, it is the case that ψ') have been explored, e.g., in Dynamic Epistemic Logic and in modal recaptures of AGM (Spohn 1988; Segerberg 1995; Lindström and Rabinowicz 1999; Board 2004; Van Ditmarsch 2005; Asheim and Sovik 2005; Leitgeb and Segerberg 2005; Van Benthem 2007; Van Ditmarsch et al. 2008; Baltag and Smets 2008b; Girard and Rott 2014).

Idea (3) is variously related to work on tautological or analytic entailment (Parry 1933, 1968; Dunn 1972; Angell 1977; Deutsch 1984; Fine 1986; Correia 2004; Ferguson 2014), topic logic (Burgess 2009), awareness logic (Fagin and Halpern 1988; Schipper 2015), and dependence or set-assignment logic (Epstein 1981, 1993). As we will see, besides being non-monotonic thanks to (2), our $X^\varphi\psi$s will turn out to be hyperintensional, differentiating between intensionally equivalent contents, thanks to their topicality filter (3).

Take a propositional language $\mathcal{L}$ with a countable set $\mathcal{L}_{AT}$ of atomic formulas, p, q, r $(p_1, p_2, ...)$, negation $\neg$, conjunction $\wedge$, disjunction $\vee$, a conditional $\prec$, X standing for a generic two-place TSIM, round parentheses as auxiliary symbols $($, $)$. I use $\varphi, \psi, \chi, ...,$ as metavariables for formulas of $\mathcal{L}$. The well-formed formulas are items in $\mathcal{L}_{AT}$ and, if φ and ψ are well-formed formulas, so are the following:

$$\neg\varphi \mid (\varphi \wedge \psi) \mid (\varphi \vee \psi) \mid (\varphi \prec \psi) \mid X^\varphi\psi$$

Outermost brackets are usually omitted. I identify $\mathcal{L}$ with the set of its well-formed formulas.

In the metalanguage I use variables $w, w_1, w_2, ...$, ranging over worlds, x, y, z ($x_1, x_2, ...$), ranging over topics, sometimes the symbols $\Rightarrow, \Leftrightarrow, \&, or, \sim, \forall, \exists$, read the usual way. A *frame* for $\mathcal{L}$ is a tuple $\mathfrak{F} = \langle W, \{R_\varphi \mid \varphi \in \mathcal{L}\}, \mathcal{T}, \oplus, t \rangle$, understood as follows:

- W is a non-empty set of possible worlds.

- $\{R_\varphi \mid \varphi \in \mathcal{L}\}$ is a set of accessibilities between worlds, indexed to formulas: each $\varphi \in \mathcal{L}$ has its own $R_\varphi \subseteq W \times W$. These may satisfy a number of different constraints, as we will see throughout the coming three chapters (in particular, we will see that the constraints introduced for some of the semantics explored there will make it so that R_φ turns out to be indexed, in fact, to the truth set of φ: the sets of worlds where φ is true).

- $\mathcal{T}$ is a non-empty set of topics: the subject matters the formulas of our language $\mathcal{L}$ can be about.

- $\oplus$ is topic fusion: an idempotent, commutative, associative binary operation on $\mathcal{T}$, familiar from chapter 2, making of topics part of larger topics. Fusion shall be, again, unrestricted, i.e., $\oplus$ is always defined on $\mathcal{T}$: $\forall xy \in \mathcal{T} \, \exists z \in \mathcal{T}(z = x \oplus y)$. Topic parthood, $\leq$, can then be defined the usual way: $x \leq y := x \oplus y = y$. Thus, it's a partial ordering again. Then $\langle \mathcal{T}, \oplus \rangle$ is a join semilattice, and $\langle \mathcal{T}, \leq \rangle$ a poset.

- t, the topic-assignment function, assigns an item in $\mathcal{T}$ to each item in $\mathcal{L}_{AT}$ and is extended to the whole of $\mathcal{L}$ as follows: if the set of atoms in φ is $\mathfrak{At}\varphi = \{p_1, ..., p_n\}$, then:

$$t(\varphi) = \oplus \mathfrak{At}\varphi = t(p_1) \oplus ... \oplus t(p_n).$$

A formula is about what its atoms, taken together, are about. As in chapter 2, this ensures the desired topic-transparency of all of our operators.

An important remark on the last point. Notice that, besides the other connectives, this set-up makes X topic-transparent, too: $t(X^\varphi\psi) = t(\varphi) \oplus t(\psi)$. I've settled for this for simplicity, but admittedly it's not quite right. Attitude ascriptions don't seem to be, in general, topic-transparent: the proposition that Mary believes that Scotland is lovely, unlike the proposition that Scotland is lovely, is about what Mary believes. This much seems clear, and if one sticks to the view, flagged in the previous chapter, that all logical vocabulary should satisfy Transparency, then one shouldn't take the TSIMs as logical vocabulary – which may be fine given that they express attitude ascriptions.

What is not clear to me is how iterated attitude ascriptions should work *qua* topicality: what's the topic of the proposition that Mary believes that she believes that Scotland is lovely? How does it relate to the topic of the proposition that Mary believes that Scotland is lovely? Currently, I have no idea.

A *model* $\mathfrak{M} = \langle W, \{R_\varphi \mid \varphi \in \mathcal{L}\}, \mathcal{T}, \oplus, t, \Vdash\rangle$ is a frame with an interpretation $\Vdash \subseteq W \times \mathcal{L}_{AT}$, relating worlds to atoms: we read '$w \Vdash p$' as meaning that p is true at w, '$w \nVdash p$' as $\sim w \Vdash p$. At each world, each atom will be either true or untrue, and not both. $\Vdash$ is extended to all formulas of $\mathcal{L}$ thus:

(S¬) $w \Vdash \neg\varphi \Leftrightarrow w \nVdash \varphi$

(S∧) $w \Vdash \varphi \wedge \psi \Leftrightarrow w \Vdash \varphi \ \& \ w \Vdash \psi$

(S∨) $w \Vdash \varphi \vee \psi \Leftrightarrow w \Vdash \varphi \ or \ w \Vdash \psi$

(S≺) $w \Vdash \varphi \prec \psi \Leftrightarrow \forall w_1(w_1 \Vdash \varphi \Rightarrow w_1 \Vdash \psi)$

(SX) $w \Vdash X^\varphi\psi \Leftrightarrow [1] \ \forall w_1(wR_\varphi w_1 \Rightarrow w_1 \Vdash \psi) \ \& \ [2] \ t(\psi) \leq t(\varphi)$

The truth conditions for $\neg$, $\wedge$, $\vee$ are the usual ones, and $\prec$ is strict implication. As for the truth clause (SX) for the TSIMs: for $X^\varphi\psi$ to come out true at w, we ask for two things to happen, flagged as [1] and [2] there. Reading them in English:

(1) ψ must be true at all worlds w_1 one looks at, via the accessibility indexed to φ. One can think of our agent as being located at w, and as accessing, given φ or via input

φ, a set of worlds where the truth of ψ is to be checked (again, more precise readings of '$wR_\varphi w_1$' will come up later, when we look at specific TSIMs). This we may call the 'strictly truth-conditional' component making of $X^\varphi \psi$ a variably strict, conditional-like quantifier over worlds.

(2) ψ must be fully on topic with respect to φ. This is the aboutness-preservation component. As anticipated above, we require full inclusion of $t(\psi)$ in $t(\varphi)$, nothing less. I'll discuss the opportunity of relaxing this – and, how to do it – for some of our topic-sensitive operators in Section 5.5.

(SX) can be equivalently expressed using set-selection functions: Lewis (1973), pp. 57-60. Each $\varphi \in \mathcal{L}$ comes with a function $f_\varphi : W \to P(W)$. One can think, again, of our agent as located at w, and f_φ will map w to the set of worlds accessible given φ, where the truth of ψ is to be checked: $f_\varphi(w) = \{w_1 \in W \mid wR_\varphi w_1\}$. If $|\varphi| = \{w \in W \mid w \Vdash \varphi\}$ (the thin proposition associated with φ, to talk Yablovian), we can compactly rephrase component [1] in the clause for X so as to get:

(SX) $w \Vdash X^\varphi \psi \Leftrightarrow$ [1] $f_\varphi(w) \subseteq |\psi|$ & [2] $t(\psi) \leq t(\varphi)$

The two formulations are equivalent, as $wR_\varphi w_1 \Leftrightarrow w_1 \in f_\varphi(w)$. However, either formulation is at times handier than the other. In particular, in Section 3.5, at the end of this chapter, I will phrase the additional constraints on the semantics of our TSIMs, which deliver the operators of chapters 4, 5, and 6, using the functions.

Finally, valid entailment is, completely standardly, truth preservation at all worlds of all models. With Σ a set of formulas:

$\Sigma \vDash \psi \Leftrightarrow$ in all models $\mathfrak{M} = \langle W, \{R_\varphi \mid \varphi \in \mathcal{L}\}, \mathcal{T}, \oplus, t, \Vdash \rangle$ and for all $w \in W$: $w \Vdash \varphi$ for all $\varphi \in \Sigma \Rightarrow w \Vdash \psi$

For single-premise entailment, I write $\varphi \vDash \psi$ for $\{\varphi\} \vDash \psi$. Validity for formulas, $\vDash \varphi$, that is, truth at all worlds of all models, is $\emptyset \vDash \varphi$, entailment by the empty set of premises.

Overall, the setting is pretty conservative and classical, logically speaking: the logic induced by the semantics for the extensional operators is just classical propositional, with $\prec$ a strict S5-like conditional (i.e., one equivalent to the necessitation of a material conditional, where the relevant necessity is S5). I take it as a selling point of the semantics, that it allows us to achieve a number of, hopefully, interesting results just by adding to this conservative setting a variably strict, two-place modal with a topicality constraint. (In various places in the book, e.g., in Section 6.3, I will, however, mention important limitations tied to the fact that the framework is *so* classical.)

I will now discuss a number of validities and invalidities concerning our TSIMs in some detail. The validities featured in this chapter involve the basic semantics I have just presented, which imposes no conditions on the accessibility relations R_φ, or functions f_φ. The featured invalidities remain invalid also in the stronger settings of the coming chapters.

3.2   Conjunctivology

The TSIMs are 'fully conjunctive': they are closed with respect to Conjunction Introduction, Elimination, and Commutation, in the following senses:

(Simplification) $X^\varphi(\psi \wedge \chi) \vDash X^\varphi\psi$ $X^\varphi(\psi \wedge \chi) \vDash X^\varphi\chi^1$

(Adjunction) $\{X^\varphi\psi, X^\varphi\chi\} \vDash X^\varphi(\psi \wedge \chi)^2$

(Commutativity) $X^\varphi(\psi \wedge \chi) \vDash X^\varphi(\chi \wedge \psi)$

Simplification has it that if (given φ) one Xs a conjunction, then as a consequence (still given φ) one Xs the conjuncts

---

[1]Though most (in)validities are easily established, I sometimes add little proofs in footnotes, just in case. For this one, *Proof:* I do the first one (for the second, replace ψ with χ appropriately). Let $w \Vdash X^\varphi(\psi \wedge \chi)$. By (SX), for all w_1 such that $wR_\varphi w_1$, $w_1 \Vdash \psi \wedge \chi$, thus by (S∧), $w_1 \Vdash \psi$. Also, $t(\psi \wedge \chi) = t(\psi) \oplus t(\chi) \leq t(\varphi)$, thus $t(\psi) \leq t(\varphi)$. Then, by (SX) again, $w \Vdash X^\varphi\psi$.

[2]*Proof:* let $w \Vdash X^\varphi\psi$ and $w \Vdash X^\varphi\chi$, that is, by (SX): for all w_1 such that $wR_\varphi w_1$, $w_1 \Vdash \psi$ and $w_1 \Vdash \chi$, so by (S∧) $w_1 \Vdash \varphi \wedge \psi$. Also, $t(\psi) \leq t(\varphi)$ and $t(\chi) \leq t(\varphi)$, thus $t(\psi) \oplus t(\chi) = t(\psi \wedge \chi) \leq t(\varphi)$. Then, by (SX) again, $w \Vdash X^\varphi(\psi \wedge \chi)$.

separately. If X is belief: one who believes (given something) that John is tall and handsome, believes (given the same things) that John is tall; if X is knowability: one who (given, etc.) would be in a position to know that the Earth is round and spinning, would be in a position to know that the Earth is round; and so on. Adjunction is the other way around: one who believes (ditto) that John is tall and that he is handsome, believes that John is tall and handsome, etc. Commutativity says that the syntactic order – the order in which the conjuncts are presented linguistically – doesn't matter.

The insight behind the TSIMs being so well-behaved with respect to conjunction involves taking the aboutness of (*de dicto*) intentionality seriously as a feature of propositional mental states. If thinking that φ has to be taken, not just as having a sentence (possibly of mentalese) tokened in the head, but as a contentful intentional state, one endowed with aboutness, then one who thinks that φ, thinks about a certain situation or issue or, well, topic towards which one's mind is directed. The mental state must be about what the proposition which makes for its content is about, and we have argued in chapter 2 for the opportunity of seeing the topic as a constituent of the (thick) proposition itself.

Simplification has it that one who thinks about the whole, thinks about the parts. You can't think that John is tall and handsome without thinking that John is tall, can you? Thinking that John is tall and handsome without thinking that he is tall would be a bit like thinking that John is tall without thinking that John is tall, wouldn't it? That's because, by thinking the whole, you have already thought about the parts: your thought must be, so to speak, directed to the whole topic of the proposition that makes for its content, and so it is already directed to the parts.

Adjunction has it that one who thinks about the parts, thinks about the whole. You can't think that John is tall and that John is handsome without thinking that John is tall and handsome because, by thinking about the parts all together, you have already thought about the whole: there's nothing more for you to do.

Commutativity has it that the order in which the parts are listed in language doesn't matter: when you think that John is tall and handsome, the content you think about is exactly the same as the content you think about when you think that John is handsome and tall, and so is the situation it directs you to. All of these are in the ballpark of what we already called, following Yablo, 'immanent closure' in previous chapters: they involve both entailment, that is, truth-preservation at all worlds of our models, and topic-preservation.

Also mental states of a putatively anarchic kind, such as imagination, seem to be closed under Conjunction Introduction, Elimination, and Commutation: try and *imagine* that it's windy and cloudy without imagining that it's cloudy and windy, or in particular that it's windy. (Some authors think that imagination constitutively involves mental imagery. If so, this may give us additional reasons for thinking that it's fully conjunctive: I'll come back to this in chapter 5.)

That Simplification, of all putative closure properties, has special plausibility for attitude ascriptions, has been argued by a number of authors. Holliday (2012), takes Simplification for knowledge operators (in the form: $K(\varphi \wedge \psi) \supset K\psi$) as a *pure* (contrast *deductive*) closure principle. Here's the difference: a deductive closure principle from $\varphi_1, ..., \varphi_n$ to ψ has it that if an agent comes to believe ψ starting from $\varphi_1, ..., \varphi_n$, by competent deduction, and all the while knowing each of $\varphi_1, ..., \varphi_n$, then the agent knows ψ. When we talk about fully realistic agents, this can, of course, always go wrong: the deduction may be too long and complex for the cognitive, computational, and memory resources of our knower. Deductions can take time, and, as mentioned, following (Williamson 2000, 282), in the introductory chapter 1, distraction or sudden death can always step in before one brings the conclusion home. By contrast, Simplification, *qua* pure closure principle, is such that 'an agent cannot know $\varphi \wedge \psi$ without knowing ψ – regardless of whether the agent came to believe ψ by "competent deduction" from $\varphi \wedge \psi$' (Holliday 2012, 15).

Williamson – a friend of full epistemic closure, unlike Holliday – has considerations which matter for Simplification in chapter 12 of *Knowledge and Its Limits*. Here, distributivity over conjunction for knowledge is key for the famous Williamsonian argument against verificationism, which resorts to Fitch's paradox. Williamson generalizes and conjectures that distributivity may hold for all positive attitudes: in conceiving a conjunction, one conceives the conjuncts; in accepting a conjunction, one accepts the conjuncts, etc. (It needn't hold for negative ones: one can *dis*believe that $\varphi \wedge \psi$ because one disbelieves φ, but doesn't thereby disbelieve ψ.)

Williamson's point is in line, again, with Yablovian immanent closure: deductive closure principles may always fail for everyday agents insofar as they involve performing some inferential action or other, for 'there is no form of inference that one can be relied on to carry out exceptionlessly' (Williamson 2000, 282). But Simplification, *qua* pure closure principle, is special:

> Knowledge of a conjunction is *already* knowledge of its conjuncts. [...] There is no obstacle here to the idea that knowing a conjunction *constitutes* knowing its conjuncts, just as, in mathematics, we may *count* a proof of a conjunction as a proof of its conjuncts, so that if $p \wedge q$ is proved then p is proved, not just provable. (Williamson 2000, 282-3)

Adjunction may look less straightforward than Simplification. Our $X^{\varphi}\psi$s, recall, look like a sort of variably strict conditionals, and Adjunction looks exactly like what is sometimes called 'Conjunction in the Consequent' in the literature on conditional logics. Now there's an old chestnut by Quine (1960) concerning same-antecedent counterfactuals not allowing for conjunction of their consequents. Quine's φ has Caesar being in command of the US troops in the Korean war. Given this, one can think that Caesar uses atomic bombs, $X^{\varphi}\psi$, if one sticks to the weapons available in the Korean war; one can think that Caesar uses catapults, $X^{\varphi}\chi$, if one sticks to the military apparatus available to Caesar. One wouldn't thereby infer $X^{\varphi}(\psi \wedge \chi)$, that one thinks that Caesar uses both nukes and catapults. Surely one *can* think that, too, but it shouldn't come as an automatic entailment mandated by the logic of intentionality.

I think, however, that several considerations speak in favour of Adjunction. First, doubts may have to do with the formalism not fully capturing the contextual character of the background information which fixes the worlds one looks at, given φ, where the relevant contexts may concern, e.g., the particular issues which are salient for the agent. Our set-selection functions f_φ are indexed only to formulas, but it seems clear that the same φ can trigger different thoughts for the same agent in different contexts. One could represent it in the formalism, if one cared, by adding to our frames a set of contexts and indexing TSIMs to them: given φ one thinks that ψ in context c_1, that χ in context c_2, and cross-contextual Adjunction will fail. The Quinean example embeds a clear shift. Once one sticks to a single context, Adjunction will work: you cannot, for instance, imagine in one go that it's cloudy and that it's windy without imagining that it's cloudy and windy, can you?

Second, further considerations may speak in favour of Adjunction, specifically in cases where the relevant mental states involve mental imagery. As already announced, I postpone them to chapter 5 when we'll meet imagination as mental simulation, an activity often – if controversially, as we will see – tied to mental imagery.

Third, some pull against Adjunction may come from the consideration of probabilistic attitudes or attitudes that need the passing of some intermediate threshold in a degree-theoretic structure in order to be triggered. Suppose we read '$X^\varphi \psi$' as 'One takes ψ as likely enough given φ', where this is understood as $p(\psi|\phi)$ (the conditional probability of ψ given φ) crossing a threshold θ, say, between $\frac{1}{2}$ and 1, for the subject. Take the Lottery Paradox (Kyburg 1961): for each ticket i, $1 \leq i \leq n$, of a large enough fair lottery L, given that L will have exactly one winner, one will find it likely enough that i will lose. However (given the same thing) one will not find it likely at all that ticket 1 will lose, and ticket 2 will lose, and ..., and ticket n will lose. Each of $X^\varphi \psi_1, ..., X^\varphi \psi_n$ will cross the threshold, but $X^\varphi(\psi_1 \wedge ... \wedge \psi_n)$ won't, and Adjunction will fail.

This does not happen, however, with $\theta = 1$. The TSIMs considered here and until chapter 7 included, besides representing all-or-nothing attitudes, should not be understood as tied to the passing of an intermediate threshold. Adjunction, then, could not fail for them on the basis of considerations of such degree-theoretic nature. (We *will* talk of a probabilistic, topic-sensitive, variably strict conditional operator in chapter 8; we'll see how things go for it with Adjunction/Conjunction in the Consequent, when we get there.)

That is not to deny that it is a deep issue, how adjunctive all-or-nothing attitudes *could* cohere with counterpart graded or probabilistic attitudes, when such natural counterparts exist. This may not be the case, e.g., for supposition: it seems that one either supposes something, or not, and there's no degrees of supposing. But it is the case, arguably, for belief.

Hannes Leitgeb (2017)'s *stability theory of belief* focuses primarily on belief for logically idealized agents. The bridge between all-or-nothing and graded belief for such agents is given by the Humean Thesis: they believe in the all-or-nothing sense a proposition iff they assign it a stably high degree of belief or subjective probability. I won't enter into the specifics of Leitgeb's notion of stability, which does deliver that all-or-nothing belief is indeed adjunctive. But I mention that the stability view saves Adjunction by introducing, again, a certain context-sensitivity: it implies a dependency of all-or-nothing belief on the epistemic context the agents find themselves in: what is salient for them, what the focus of their attention is (Leitgeb 2017, Section 3.3). Roughly: all-or-nothing belief can be adjunctive in spite of being tied to intermediate degree-of-belief thresholds θ, insofar as the value of θ can change across contexts, and what is going on in Lottery Paradox cases are switches of epistemic context. (There are interesting connections between Leitgeb's view and my own, which will be mentioned in Section 6.2.)

Finally, some intuitive pull against Adjunction may come from the consideration of Preface Paradox cases (Makinson 1965), of which some philosophers think that they are different enough from Lottery Paradox cases (Foley 1993; Pollock 1994). Something will be said about this in the epistemic TSIM setting of the coming chapter: as we will see in Section 4.3 there, TSIMs may be able to handle cases which look Preface-Paradox-like via the invalidity of a principle closely related to, but not quite the same as, Adjunction.

Commutativity, which follows trivially from Simplification and Adjunction, can prompt a discussion of order-of-
presentation effects in language and thought. We get a long
list of search results on Amazon and we stop when we find
an article we judge satisfactory. Had the same items in the
list been arranged differently, with the article further down,
we may never have bought it. If we take the list as a long
conjunction of sentences, $\varphi \wedge \psi \wedge \chi \wedge ...$, order matters (Schipper
2015, 83).

But then, either '$\wedge$' is no Boolean conjunction here because
it's not commutative or one has failed to properly parse the
syntax of the sentence, or to compute its meaning from that
of its constituents so as to get to its propositional content.
Surely 'a computer program that can determine whether
$\varphi \wedge \psi$ follows from some initial premises in time τ might
not be able to determine whether $\psi \wedge \varphi$ follows from those
premises in time τ' (Fagin and Halpern 1988, 53, notation
modified). However – again, once we factor out cognitive and
computational limitations of the kind that may show up while
parsing or manipulating the syntax of sentences – this is no
difference concerning the scenario towards which the mind of
an intentional agent is directed.

To be sure, we may want to model agents whose attitudes
display some closure failure due to issues connected to
the parsing of syntax. It's not clear that this should or
could easily be the business of a properly model-theoretic
semantics – though there's some hope, I think, if one resorts
to logics in the ballpark of the Lambek calculus, which are
in a way syntax-sensitive by representing linguistic string
concatenation, and which can be provided with 'linguistic
frame semantics' in the general framework of substructural
logics: see Restall (1999), 307-8. Levesque (1984)'s classic
remark on merely syntactic approaches to knowledge and
belief, where one's epistemic or doxastic state is represented
by a set of formulas which can fail closure under any non-
trivial notion of logical consequence, hinges on this. He makes
the point with respect to belief and disjunction:

> The syntactic approach [...] is too *fine-grained* in
> that it considers any two sets of sentences as dis-

> tinct semantic entities and, consequently, different belief sets. Consider, for example, the disjunction of ϕ and ψ. There is no reason to suppose that $B(\phi \vee \psi) \equiv B(\psi \vee \phi)$ would be *valid* given a syntactic understanding of B since $\phi \vee \psi$ may be in the belief set while $\psi \vee \phi$ may not. [But] if we consider intuitively what '*It is believed that either ϕ or ψ is true*' is saying, the order seems to be completely irrelevant [...] $\phi \vee \psi$ is believed iff $\psi \vee \phi$ is because these are two lexical notations for the *same* belief. (Levesque 1984, 199-201, notation modified)

Talk about disjunction brings us to the next Section, where the interaction of our TSIMs with disjunction is seen to be very different from that with conjunction. We stop talking about valid entailments, and we start exploring invalidities.

3.3   Disjunctivology

We think about things vaguely, without this entailing that we think about vague things. It happens with intentional states involving mental imagery: when one pictorially imagines that the Autumn leaves float around in St Andrews while the waves of the Northern sea crash against its high cliffs, one doesn't pictorially imagine all the details, but one wants the details to be there, so to speak. Although one doesn't imagine the town building by building, St Andrews is not a vague object in the scenario – one with an objectively indeterminate number of buildings. Either the number of buildings of St Andrews is odd or it is even. But one does not imagine it either way.

Under-determination is not tied to mental imagery, anyway. Thinking about the initial values in a mathematical sequence, one can believe, with no mental imagery involved, that either the sequence has a limit or it's divergent, but one has no view on which one it is. We need, and we get, a failure of:

(Distribution) $X^\varphi(\psi \vee \chi) \nvDash X^\varphi\psi \vee X^\varphi\chi$[3]

The invalidity says that when (given φ) one Xs a disjunction, this doesn't logically entail that (still given φ) one Xs either disjunct in particular. This is delivered by the fact that input φ generally has one look at a plurality of worlds. In terms of set-selection functions: f_φ can output a bunch of worlds, some of which make ψ false, some make χ false, although all make $\psi \vee \chi$ true. One reads some pages of *The Lord of the Rings* and thinks that Boromir is either left-handed or right-handed (or ambidextrous): he's a normally endowed human being, after all. One doesn't thereby think that he is left-handed, and one doesn't thereby think that he is right-handed (say: *The Lord of the Rings* is telling one nothing about Boromir's dominant hand). There'll be worlds compatible with what one thinks, where Boromir is left-handed, and compatible worlds where he's right-handed instead.

Another disjunction-involving invalidity needs more justification. It may be, for some, the single hardest thing to swallow concerning our TSIMs' (non-)closure features:

(Addition) $X^\varphi\psi \nvDash X^\varphi(\psi \vee \chi)$[4]

When (given φ) one Xs that ψ, this doesn't entail that (still given φ) one Xs an arbitrary disjunction with ψ as one disjunct. This is provided by topic-sensitivity: although, in our basic semantics above, $\psi \vDash \psi \vee \chi$, Disjunction Introduction can bring in extra subject matter, delivered by χ. Going back to the terminology of chapter 2: a disjunction can say less than either disjunct insofar as it's entailed by both, but it can say less *about more*.

What's the point of representing thinkers whose intentional states can fail to be additive? Disjunction Introduction is as

---

[3] *Countermodel:* let $W = \{w, w_1, w_2\}$, wR_pw_1, wR_pw_2, $w_1 \Vdash q$ but $w_1 \nVdash r$, $w_2 \Vdash r$ but $w_2 \nVdash q$, $t(p) = t(q) = t(r)$. Then by (S$\vee$), $w_1 \Vdash q \vee r$ and $w_2 \Vdash q \vee r$, so for all w_x such that wR_pw_x, $w_x \Vdash q \vee r$. Also, $t(q \vee r) = t(q) \oplus t(r) \leq t(p)$, thus by (SX), $w \Vdash X^p(q \vee r)$. However, $w \nVdash X^pq$ and $w \nVdash X^pr$ for both q and r fail at some R_p-accessible world. Thus by (S$\vee$), $w \nVdash X^pq \vee X^pr$.

[4] *Countermodel:* let $W = \{w, w_1\}$, wR_pw_1, $w_1 \Vdash q$, $t(r) \nleq t(p) = t(q)$. Then $t(q) \leq t(p)$, so by (SX), $w \Vdash X^pq$. But $t(q \vee r) = t(q) \oplus t(r) \nleq t(p)$, thus $w \nVdash X^p(q \vee r)$.

basic an inference as Conjunction Elimination, one might say. If the goal is to represent realistic agents who can fail to perform even the simplest inferences, whence the asymmetry between conjunction and disjunction?

But the asymmetry is motivated by the difference in topic-preservation: to think that $\varphi \wedge \psi$, you have to think both about what φ is about, and about what ψ is about. Instead, you can think that φ without thinking about what $\varphi \vee \psi$ is about. You may not even be in a position to think about the latter, because you have no way to think about what ψ is about, even if you are a perfect reasoner: the topic of ψ may be *alien* to you.

One may want to model, that is, agents that, in spite of being deductively unbounded, are *conceptually* bounded or limited: they are not on top of certain subject matters, because they lack the concepts required to grasp them. I exploit Williamson again:

> $\wedge$ elimination has a special status. It may be brought out by a comparison with the equally canonical $\vee$-introduction inference to the disjunction $p \vee q$ from the disjunct p or from the disjunct q. Although the validity of $\vee$-introduction is closely tied to the meaning of $\vee$, a perfect logician who knows p may lack the empirical concepts to grasp (understand) the other disjunct q. Since knowing a proposition involves grasping it, and grasping a complex proposition involves grasping its constituents, such a logician is in no position to grasp $p \vee q$, and therefore does not know $p \vee q$. In contrast, those who know a conjunction grasp its conjunct, for they grasp the conjunction. (Williamson 2000, 282-3)

Williamson makes the point with respect to knowledge, but it easily generalizes to a variety of attitudes: one cannot believe, or suppose, or imagine, or ..., that φ, if one doesn't grasp the proposition that φ. Williamson may or may not want topics to be constituents of content (my bet: he will not); but I like the idea, as made apparent in chapter 2. One may fail to grasp a (thick) proposition because one lacks the concepts needed to grasp its subject matter. (And, if the key 2C claim from that chapter is right, one may not be in a position to recover the subject matter even when one is perfectly on top of truth conditions.) To adapt an example from (Stalnaker 1984, 88): William III may have known (or, believed, etc.) that England could avoid war with France, without thereby knowing (or, ditto) that either England could avoid war with France, or France could develop a nuclear arsenal: he had no idea what nuclear weapons might be, so he could not entertain nuclear-weapons-involving thoughts.

Addition can fail due to alien topics. We may want it to fail also to model agents who disregard *irrelevant* topics, even when they are not alien to them. This will matter especially in chapter 5, when we talk about imagination and suppositional thinking, but I can anticipate a bit here.

What if Stauffenberg had put the bomb on the other side of the table? Would that have killed the infamous dictator? When the suppositional input is $\varphi = $ *Stauffenberg puts the bomb on the other side of the table*, and one integrates the input with what one knows or believes concerning the involved bit of history, etc., one may conclude by imagining that $\psi = $ *Hitler gets killed*. We may not want our agent to automatically imagine that either Hitler gets killed or there's life on Kepler-442b, even when they are on top of the concepts needed to grasp the second disjunct. Remember how topics or subject matters, as we have seen in chapter 2, naturally connect to issues or questions. The connection explains our resistance to that automatic entailment: it would bring too much of a gratuitous departure from the underlying issue, or question one is addressing, when one engages in the imaginative exercise: What happens to Hitler?

One may think that the full inclusion of $t(\psi)$ in $t(\varphi)$ is too draconian a requirement for TSIM '$X^{\varphi}\psi$' to represent imagination as mental simulation: there's no way to have both a natural account of topicality, and such 'analytic' inclusion. After all, how would the topic of the proposition that Stauffenberg puts the bomb on the other side of the table naturally include that of the proposition that Hitler gets killed? Only the latter is about Hitler, right? Imagination as mental simulation may be somewhat regimented, but it's not *so* regimented never to expand the topic of the initial supposition. I'll get back to this issue at the end of chapter 5.

There are yet further reasons for wanting Addition to fail for certain mental states, anyway. $\psi \vee \neg\chi$ is equivalent to $\neg(\neg\psi \wedge \chi)$ in both truth conditions and topic: they hold at the same possible worlds, and they are about the same things. If $X^\varphi\psi$ is to stand for some knowledge operator (as it will, in chapter 4: 'One is in a position to know ψ, given information φ'), Addition would give us that one who is in a position to know ψ (and who is on top of the relevant concepts) is automatically in a position to know $\psi \vee \neg\chi$. This is equivalent to being in a position to know $\neg(\neg\psi \wedge \chi)$.

But this is wrong for most fallibilists about knowledge (Brown 2018). Typical fallibilist examples: I know I have hands, but I'm not in a position to know that it's false that I am a handless brain in a vat. I know this is a zebra, but I'm not in a position to know that it's false that it's no zebra but a cleverly disguised mule. For fallibilists, this can hold even when I am, again, a perfect reasoner who never fails to put two and two together, logically speaking; and moreover, I am perfectly on top of all the relevant topics: I don't lack concepts needed to think about envatted brains, or cleverly disguised mules. The TSIM-knowability operator explored in chapter 4 will represent such fallibilist-friendly insights (not only, but also) by failing Addition.

Our two-place TSIMs look like a sort of variably strict conditionals, I said. One easily thinks of their topic-inclusion requirement as making of them a sort of especially demanding *relevant* conditionals. Relevant logics aim at capturing a notion of conditionality free from the so-called paradoxes of the material and strict conditional, such as (let's use '$\rightarrow$', just for this Section) $\varphi \rightarrow (\psi \rightarrow \psi)$, $\varphi \rightarrow (\psi \vee \neg\psi)$, and $(\varphi \wedge \neg\varphi) \rightarrow \psi$. Anderson and Belnap (1975) held that a formula of the form $\varphi \rightarrow \psi$ should only be a theorem, or a logical validity, if antecedent and consequent share an atomic formula. This was called the Variable Sharing Property (VSP) (Dunn and Restall 2002, 27).

The VSP captures in syntactic fashion the idea that there must be overlap between the antecedent and consequent of a logically good conditional. Because the VSP only demands overlap, relevant conditionals are generally additive: if $\varphi \to \psi$ satisfies the VSP, so does $\varphi \to (\psi \vee \chi)$. However, there's a little-known sub-family of relevant logics, called conceptivist or containment logics, which fail Addition, and whose story is masterfully reconstructed by Ferguson (2014). Briefly put: conceptivist logics look essentially like ordinary modal logics, featuring conditional operators with (i) the truth conditions of a strict conditional, plus (ii) a syntactic relevance filter on top of it: for $\varphi \to \psi$ to be logically valid, one requires all the atomic formulas in ψ to occur in φ. The TSIMs' similarity to such non-additive conditionals is patent.

Now mainstream relevant logicians like Richard Routley-Sylvan criticized conceptivist logics as gerrymandered, because of such (i)-(ii) two-component nature and because of the syntactic nature of filter (ii) (Ferguson 2014, 335-6). The problems arising from the irrelevance of ordinary classical or modal entailments 'are not repaired simply by throwing on a variable-inclusion filter' (Routley 1982, 100). However, I don't think considerations of this kind can point at any trouble for our TSIMs. (Ferguson already makes a good case for conceptivist logics in general; see also Ciuni et al. (2018).)

First, I think the case for the usefulness of TSIM theory and, in general, of topic-sensitive intentional operators, is a holistic one. The core idea is that the logic of thought must be topic-sensitive, because what ψs one thinks given that one thinks that φ must depend, not only on the conditions under which φ and ψ are true, but also on what they are about. One judges the idea on the basis of the things one can do with it – and I hope the book as a whole can convince its reader that these are numerous and interesting.

Second, unlike (ii) above, the topic-inclusion constraint in the semantics of our TSIMs is no syntactic filter: it is fully semantic. This marks a difference with various kinds of awareness logics (Fagin and Halpern 1988; Velázquez-Quesada 2011, 2014; Schipper 2015), which have been used to model non-omniscient agents in doxastic-epistemic logic. In such approaches, unawareness is generally understood as lack of conception, rather than lack of information (Schipper 2015, 79-80). In the seminal work of Fagin and Halpern (1988), awareness is represented syntactically: one is aware of φ

when φ belongs to a set of formulas, the agent's awareness set $\mathcal{A}$. Next, there is a distinction between *implicit* and *explicit* knowledge or belief (this will be discussed in Section 7.3). The former gets the usual Hintikkan characterization, whereas explicit attitudes are defined as the combination of the implicit ones with awareness, which acts as a filter on the implicit: an agent has the explicit attitude towards φ when it has the implicit one and, in addition, $\varphi \in \mathcal{A}$. The closest variant to our topic-sensitive approach is 'propositionally determined awareness' (see Halpern (2001), 327, focusing on knowledge): one is aware of φ just in case one is aware of all of φ's atomic formulas taken together. Because of such appeal to formulas and syntactic filters, awareness has been claimed to mix syntax and semantics (Konolige 1986).

On the contrary, the requirement for $X^\varphi \psi$ to hold is not that the atomic formulas in ψ all show up in φ. The inclusion is between topics. Whatever these might be, metaphysically speaking, they are not going to be syntactic objects. Notice that the semantics works in such a way that nothing forbids $t(p) = t(q)$ for distinct p and q: different atoms can be assigned the same topic – naturally enough, because distinct logically atomic sentences can speak about the same thing.

Third, here's one point where the key 2C claim from the previous chapter, if true, gives us an additional story against the gerrymandering charge. I have claimed since chapter 1 that topic-sensitive logics may be a useful tool even if truth conditions are ultimately reducible to topics or vice versa, against 2C. If 2C is right, though, it's not just useful to represent two components separately: they *are* separate. The two-component semantics of our intentional operators, featuring a quantification over worlds conjoined with a topic-inclusion requirement, carves at the natural joints of content.

3.4   Non-Monotonicity, Hyperintensionality

As already anticipated in Section 3.1, our two-place TSIMs are non-monotonic:

(Monotonicity) $X^\varphi \psi \nvDash X^{\varphi \wedge \chi} \psi$[5]

Although one thinks that ψ given φ, one may not think that ψ when more is added to φ. For a familiar example from the literature on non-monotonic reasoning: given that Tweety is a bird, one thinks that it flies; given that Tweety is a penguin bird, one doesn't. Again, this failure plausibly holds across a range of attitude ascriptions X can stand for. If X is conditional belief: one may believe that John is a lawyer, given the information that he studied law; one may not believe that much, given the information that John studied law but dropped out of college before getting his degree. If X is imagination as mental simulation: supposing tomorrow will be sunny, one imagines that one will play football. Supposing tomorrow will be sunny and one breaks one's leg, one won't imagine that one will play football. And so on.

Topicality is preserved here, for in general if the topic of ψ is included in that of φ, $t(\psi) \leq t(\varphi)$, then the former will also be included in a topic including the latter: $t(\psi) \leq t(\varphi) \oplus t(\chi) = t(\varphi \wedge \chi)$. The failure of Monotonicity comes from the TSIMs' variable strictness: $f_\varphi(w)$ can differ from $f_{\varphi \wedge \chi}(w)$, that is, the scenarios the agent (located at w) looks at given φ can differ from the scenarios the agent (ditto) looks at given $\varphi \wedge \chi$.

The two-place TSIMs are hyperintensional in that replacement of intensional equivalents both in their first and in their second argument can fail to preserve truth value. This feature will turn out to be pervasively important; and it's here that topic-sensitivity really does all the lifting. The hyperintensional features of our two-place TSIMs come from their topicality constraint. Remove that, and they will go back to behaving just as some kind of non-hyperintensional, variably strict conditional.

In particular, Closure under Strict Implication fails:

(Closure under $\prec$) $\{X^\varphi \psi, \psi \prec \chi\} \nvDash X^\varphi \chi$[6]

---

[5] *Countermodel:* let $W = \{w, w_1\}$, w R_p-accesses nothing, $w R_{p \wedge r} w_1$, $w_1 \nVdash q$, $t(p) = t(q) = t(r)$. Then $w \Vdash X^p q$, but $w \nVdash X^{p \wedge r} q$.

[6] *Countermodel:* let $W = \{w, w_1\}$, $w R_p w_1$, $w \nVdash q$, $w_1 \Vdash q$, $w_1 \Vdash r$, $t(r) \nleq t(p) = t(q)$. Then $f_p(w) \subseteq |q|$ and $t(q) \leq t(p)$, thus by (SX), $w \Vdash X^p q$. Also, $|q| \subseteq |r|$, thus by (S$\prec$), $w \Vdash q \prec r$. But although $f_p(w) \subseteq |r|$, $t(r) \nleq t(p)$, thus $w \nVdash X^p r$.

Even if there's no way for ψ to be true without χ being true, i.e., any ψ-world is a χ-world, one can think that ψ without thinking that χ. Such failure comes from the fact that strict implication can take one off-topic, even though all the ψ-worlds are χ-worlds, thus all the φ-selected ψ-worlds are χ-worlds. In particular, $\varphi \prec \psi$ does not entail $X^\varphi \psi$ because it may happen that $t(\psi) \not\sqsubseteq t(\varphi)$: ψ brings in extra subject matter with respect to φ.

In the same ballpark, we have:

$$(\text{Equivalence}) \quad \{X^\varphi \psi, \varphi \succ\!\!\prec \chi\} \not\models X^\chi \psi$$

(where '$\varphi \succ\!\!\prec \chi$' abbreviates '$(\varphi \prec \chi) \wedge (\chi \prec \varphi)$': φ and χ are co-intensional, or co-necessary). If φ and χ are co-necessary, but topic-divergent, ψ may be fully on topic with respect to φ but not with respect to χ. Non-Monotonicity, failure of Closure under $\prec$, and Equivalence failure, have been singled out here because they will play important roles for specific TSIMs in the coming chapters. Various examples of why such failures are plausible for a variety of attitudes will be introduced and commented on as we progress through them.

A two-place hyperintensional operator and, in particular, one failing Equivalence is in for a rather weak logic. When we find operators approximately of the kind 'Given φ, ψ' (e.g. in probability logic: 'ψ is likely given φ', Adams (1998); or in dynamic belief revision: 'After revision of one's beliefs by φ, ψ is the case', Van Benthem (2007); or in conditional logic: 'If φ is/was/had been the case, then ψ is/would be/would have been the case'), they nearly always satisfy the principle whereby 'Given φ, ψ' entails 'Given χ, ψ' when φ and χ are co-necessities or intensional equivalents. E.g., we find it, under the label of Left Logical Equivalence, in the mainstream system P of non-monotonic logic, often taken as a set of minimal principles any non-monotonic logic should obey (Kraus et al. 1990). However, specific TSIMs may satisfy some *restricted* equivalence principle. This issue will keep us busy in the coming chapters. In order to get there, let us move on to ways of strengthening the rather weak basic TSIM setting.

3.5   Adding Constraints

One can impose various conditions or constraints on the accessibility relations R_φ, featuring in the semantic clause (SX) for our TSIMs we met above. The constraints are familiar from the literature on conditional logics (Nute 1984; Priest 2008). In the $X^\varphi\psi$ setting, the idea is that they give us more information on which worlds our thinker looks at given φ. They are expressed in a compact way in terms of the corresponding f_φ:

(C0) $|\varphi| \subseteq f_\varphi(w)$

(C1) $f_\varphi(w) \subseteq |\varphi|$

(C2) $f_\varphi(w) \subseteq |\psi|$ & $f_\psi(w) \subseteq |\varphi| \Rightarrow f_\varphi(w) = f_\psi(w)$

(C3) $|\varphi| \neq \emptyset \Rightarrow f_\varphi(w) \neq \emptyset$

(C4) $f_\varphi(w) \cap |\psi| \neq \emptyset \Rightarrow f_{\varphi\wedge\psi}(w) \subseteq f_\varphi(w)$

To give these an intuitive reading in English: (C0) says that all the worlds where φ is true, i.e., those in the truth set $|\varphi| \subseteq W$, are selected for the agent (located at w) to look at, given input φ. (C1) is the other way around: all the worlds selected given input φ for the agent (ditto) to look at, are worlds where φ is true. (C2) – a constraint sometimes called 'Uniformity' in the literature on conditionals: see appendix B to Starr (2019)– says that if all the worlds selected given φ make ψ true, and vice versa, all the worlds selected given ψ make φ true, then the φ-selected worlds are the same as the ψ-selected worlds. (C3) says that there'll be some world selected for the agent to look at, given input φ, if it is possibly true that φ. (C4) says that all the φ-and-ψ-selected worlds will be φ-selected worlds when the latter are compatible with the truth of ψ.

The three readings of our two-place TSIMs to be explored in the three following chapters come, respectively, from: (i) adding (C0), which gives us Knowability Relative to Information; (ii) adding (C1) (and, tentatively, (C2)), which gives us imagination as Reality-Oriented Mental Simulation; and (iii) imposing a total preordering of worlds, interpreted as representing comparative plausibility, which automatically validates (C1)-(C4) and gives us hyperintensional conditional belief and belief revision. In each case, we restrict our attention to models that satisfy the relevant constraints. Condition (C2) yields the restricted equivalence principle hinted at, just at the end of the previous Section. It will be in focus in our chapter 5, devoted to suppositional thinking, and it will pop up again in chapter 6.

3.6   Chapter Summary

This chapter has introduced two-place TSIMs: variably strict, topic-sensitive modal operators of the form '$X^\varphi\psi$', which are to represent attitude ascriptions. It has presented a basic possible worlds semantics for a simple propositional language including them. The key clause, giving the truth conditions for the TSIMs, has it that, for $X^\varphi\psi$ to be true, we require (1) that ψ be true throughout a set of worlds, selected via an accessibility relation or function indexed to φ; and (2) that ψ be fully on-topic with respect to φ. The semantics validates various inferences involving conjunction; invalidates various inferences involving disjunction; and makes the TSIM operators non-monotonic and hyperintensional. Such validities and invalidities have been defended for interpretations of X that range across a variety of attitudes. Finally, the chapter has introduced a number of constraints on the selection functions. These will give us stronger logics for the $X^\varphi\psi$s as well as different plausible readings for them, to be explored in the three following chapters.

4

Epistemic Closure, Dogmatism, Scepticism, Fallibilism

Co-authored with Peter Hawke

Constraint (C0) from Section 3.5, $|\varphi| \subseteq f_\varphi(w)$, says that all the φ-worlds are φ-selected, but allows for φ-selected worlds which are not φ-worlds. Read it as saying that no situation where φ is true is epistemically ruled out, based on the information that φ. Read the corresponding $wR_\varphi w_1$ as saying that w_1 is epistemically accessible from w, given the information that φ. We then relabel '$X^\varphi \psi$' as '$K^\varphi \psi$', for we take it as expressing *Knowability Relative to Information* (KRI): 'Given total (empirical) information φ, one would be in a position to know ψ'.

The idea comes from Dretske (1999): knowledge depends on the available (empirical) information, insofar as information narrows down the set of epistemically viable alternatives. This much seems uncontentious, but what is information here? How does the φ in '$K^\varphi \psi$' behave? Well, this is for the rest of the chapter to unpack, but to anticipate a bit here, we will follow a certain path in the taxonomy described by Floridi (2019) for this polymorphic concept: information, of the kind we work with here, is contentful, propositional (and so, we add, topic-sensitive). But, as we will see, in the KRI setting it's not factive: we depart from Dretske himself (and Floridi) in not taking all information to be perforce veridical. Although the key epistemic role of information is to act as

Topics of Thought: The Logic of Knowledge, Belief, Imagination.
Francesco Berto, Oxford University Press. © Francesco Berto 2022.
DOI: 10.1093/oso/9780192857491.003.0004

evidence carrier, and so we will often speak about evidence in this chapter, we wouldn't want to commit to the TSIM explored here representing just knowability given *evidence*. Some epistemologists argue that the latter must be factive (perhaps for it is knowledge already, as famously argued by Williamson (2000)), on pain of psychologizing the notion in an unattractive way. Others disagree with sophisticated arguments, e.g., Brown (2018). We maintain neutrality on this: if evidence turns out to be factive whereas information isn't, then information just may, on occasion, fail to carry evidence.

The standard Hintikkan framework already embeds the impulse to parameterize knowledge to information: it models one's epistemic situation as a set of worlds, understood as giving one's epistemic state. Ascriptions '$K_a\varphi$' can then be understood as capturing what is knowable for agent a on this basis. The purely descriptive interpretation in sect. 2.10 of Hintikka (1962) is: 'It follows from what agent a knows that φ'.

But, the Hintikkan framework is characteristically monotonic and, as we know, it delivers full epistemic closure and so it cannot do some of the KRI work described in this chapter. In particular, the non-monotonicity of KRI *qua* TSIM can model the defeasibility of knowledge in the following sense: sometimes more information, even when truthful, can reduce what one is in a position to know. Its topic-sensitivity invalidates controversial forms of epistemic closure while validating less controversial ones.

We have a case study demonstrating the usefulness of KRI: the Kripke-Harman dogmatism paradox. We introduce it in Section 4.1. We show that it actually embeds two subparadoxes: one can be dealt with via non-monotonicity; the other can be dealt with via topic-sensitive closure failure. In Section 4.2, we show that the KRI setting captures the factivity of knowledge without committing to the factivity of information. In Section 4.3 we speak further in defence of Adjunction, by arguing that putative counterexamples in the ballpark of the Preface Paradox may actually illustrate the failure of a less wise adjunctive principle.

Defenders of Monotonicity, or of full epistemic closure,

often emphasize that deduction must preserve knowledge and that knowledge, as per the venerable Platonic idea of *epistéme*, must rest on conclusive grounds. But the KRI framework satisfies a stability principle (Section 4.4) and a closure principle (Section 4.5) which, we submit, should be accepted by all hands in the debate. Monotonicity and closure discontents needn't reject those insights – only certain formulations thereof. Finally, in Section 4.6 we examine three invalidities making of the KRI TSIM a peculiar non-monotonic operator. We argue that the KRI setting is right in making the invalidities invalid.

4.1   The Kripke-Harman Dogmatism Paradox

The paradox is due to Kripke (2011b); but it first appeared in Harman (1973, ch. 9, sect. 2), who was reporting Kripke's ideas. It has been widely discussed (Sorensen 1988; Lasonen-Aarnio 2014; Sosa 2017). It applies to ideally astute logicians who believe all the logical consequences of what they know, on the basis of what they know. What makes the paradox a natural case study for KRI is that reactions to the paradox often reject either full closure under entailment for knowability ascriptions, or Monotonicity. As we already know, both full closure and Monotonicity naturally fail in the TSIM setting. We will show that both failures are needed to block the paradox.

Suppose P, E, and R are true (in the context of this discussion we often use capital letters, not for propositions, but as signposts for specific sentences. The notation shouldn't confuse anyone). R is the claim that E is a good Reason to think that P is false (given one's information). M is the claim that E is Misleading information on the question of whether P. The following seems clearly true, given what it is for a body of information to be misleading:

1. If P is true, and R is true (that is: E is a good reason to think that P is false), then it must be that if E is true, then M is true (that is: E is misleading information on

the question of whether P). Shortly: $(P \wedge R) \prec (E \supset M)$.[1]

Suppose one knows that $P \wedge R$ at time t_0, on the basis of information I_1. If what one is in a position to know is closed under strict implication, we get:

2. At t_0, one is in a position to know that $E \supset M$.

Suppose one comes to know E at time t_1 on the basis of new information I_2. Presumably, one's information is now: $I_1 \wedge I_2$. If what one is in a position to know grows monotonically, we get:

3. At t_1, one is in a position to know that $E \supset M$.

Since one also knows E at t_1, given that what one's in a position to know is closed under Modus Ponens, we get:

4. At t_1, one is in a position to know that M.

If one knows that E is misleading, then presumably one is rational, in the face of E, to continue believing P, ignoring the 'usual implications' of E. With no loss of generality, this gives us a dogmatism principle, as stressed by Kripke (2011b): knowing agents are immune to rational persuasion with new evidence (recall that E is stipulated to be true, although misleading, information).

This sounds bad. It is well-known that Kripke first proved certain results in modal logic. Suppose one comes across a letter to Nozick, signed by Kripke, in which Kripke claims to have plagiarized the results. The contents of the letter are false: it's just a private joke between Kripke and Nozick, but one is unaware of this. Becoming informed about the existence of such a letter undermines one's rational belief, and thereby one's knowledge (given that the latter requires the former), that Kripke produced the results. The reasoning from (1) to (4) seems to advocate that one can and should

---

[1]Recall our $\mathcal{L}$ from the previous chapter, with its semantics. $\prec$ is strict implication: there's no possible world where the antecedent of a true strict implication claim is true and the consequent false. $\supset$ is the material conditional, defined the usual way.

resist this change in belief, since one knows that the new information is misleading about Kripke's accomplishments. But it is precisely the fact that one does not know this, that fuels a rational loss of belief.

Suppose we accept the conclusion of the paradox. Still, our ordinary (purported) claims to knowledge can obviously be challenged with new counter-evidence. Thus these claims must be, on reflection, false. Scepticism looms. Alternatively, we need to defy the reasoning that leads to the paradoxical conclusion.

One way targets Monotonicity (Harman 1973, Ch. 9, Sect. 2). Take the step from (2) to (3): if $E \supset M$ is knowable at t_0, then it remains knowable at t_1 if the only change for the agent is that they have received new information. To abandon Monotonicity is to allow that the additional incoming information I_2 might reduce what one is in a position to know. Using our KRI notation, one might accept counter-instances to Monotonicity of the form:

$$(I_1 \wedge I_2) \prec I_1$$

$$K^{I_1}(E \supset M)$$

$$\neg K^{I_1 \wedge I_2}(E \supset M)$$

Even if one is in a position to know that $E \supset M$ on the basis of information I_1, one may not be so positioned when I_2 is added.

Another way targets Closure under strict implication(Sharon and Spectre 2010, 2017) . Using our KRI notation, one might accept counter-instances to Closure under $\prec$ of the form:

$$K^{I_1}(P \wedge R)$$

$$(P \wedge R) \prec (E \supset M)$$

$$\neg K^{I_1}(E \supset M)$$

Even if one is in a position (on the basis of I_1) to know that $P \wedge R$, and this strictly implies that $E \supset M$, one is not in a position (ditto) to know the latter.

Closure under $\prec$ may be found by some more plausible than Monotonicity, so it is worth bolstering the appeal of this second way. Harman's solution concedes that (2) holds. Thus, at t_0, one knows that any counter-evidence to P that one might receive is guaranteed to be misleading. We can accept, with Harman, that if actually presented with new counter-evidence then one would be rationally swayed and lose some knowledge. A residual paradox remains: at t_0, one would be rational to do everything one can to avoid any possible counter-evidence, especially if one knows that it will hold one under its sway if it appears. Kripke (2011b) points out that this is equally bad dogmatism: how can rational agents be entitled to actively avoid any source of information challenging whatever they take to constitute their knowledge? A similar point is made by Sharon and Spectre (2010), 310-11. Dropping Closure under $\prec$, thereby allowing for knowing agents that are receptive to counter-evidence, is still appealing.

But then, the dogmatism paradox embeds *two* sub-paradoxes: one based on Monotonicity, one on Closure. Here's the structure of the paradoxical reasoning in the KRI notation:

5. $P \prec \neg(E \land \neg P)$ (by classical propositional and modal logic)

6. $K^{I_1} P$ (premise)

7. $K^{I_1} \neg(E \land \neg P)$ (by (5), (6), and Closure under $\prec$)

8. $K^{I_1 \land I_2} E$ (premise)

9. $K^{I_1 \land I_2} \neg(E \land \neg P)$ (by (7) and Monotonicity)

10. $K^{I_1 \land I_2}(E \land \neg(E \land \neg P))$ (by (8), (9) and Adjunction)

11. $(E \land \neg(E \land \neg P)) \prec P$ (by classical propositional and modal logic)

12. $K^{I_1 \land I_2} P$ (by (10), (11) and Closure under $\prec$)

To discern the stakes, again interpret E as a claim that inductively supports $\neg P$. We now use $\neg(E \land \neg P)$ to capture the idea that E is misleading if E and P are true. This technique for formalizing *misleading evidence* has proven useful in mainstream epistemology: see, for instance, Vogel

(2014). (12) captures a significant element of the paradoxical reasoning: new information cannot carry counter-evidence that undermines previous knowledge, since an agent knows that any counter-evidence is misleading: see (10).

But to achieve a paradox using Monotonicity, the intervening steps from (7) to (11) are inessential. Our first sub-paradox:

13. $K^{I_1}P$ (premise)

14. $K^{I_1 \wedge I_2}P$ (by Monotonicity from (13))

Putting aside memory failure, information seems cumulative: new information can only tell one more about the world. But the example of losing one's knowledge of the genesis of Kripke's theorem through the misleading letter bears directly on the reasoning from (13) to (14), and so on Monotonicity. In this case, knowledge can be lost with the accrual of novel knowledge-producing information, since that information undermines formerly rational beliefs, and so knowledge resting on those beliefs.

Next, (5), (6), and (7) provide a Closure-based sub-paradox:

15. $P \prec \neg(E \wedge \neg P)$ (premise)

16. $K^I P$ (premise)

17. $K^I \neg(E \wedge \neg P)$ (by (15), (16), and Closure under $\prec$)

This is independently bad. For emphasis, set E to be I. Then $\neg(I \wedge \neg P)$ says that the agent's total information I is not misleading on the question of P. But then (15), (16) and (17) seem to say: if one knows anything, one is positioned to know that one's total information is never misleading. Isn't it objectionably circular to claim that one's total information gives assurance that one's total information is never misleading?

Luckily, our KRI TSIM invalidates both Monotonicity and Closure under strict implication! Let's explore its workings in more detail.

4.2   Information, Factivity

Remember the twofold truth conditions for our TSIMs from Section 3.1 in the previous chapter: in the KRI setting, for $K^\varphi \psi$ to come out true we ask (1) that ψ be true throughout a set of worlds picked out by f_φ; (2) that ψ be fully on topic with respect to φ. Knowability is determined by the available information φ twice over: via the worlds it allows one to access epistemically, and via the topic it concerns.

This complies with insights about informativeness, and its relation to knowledge. Dretske (1999) has it that knowledge depends on information: to learn that Beth's grandmother is ill, one requires information to that effect. Information should not be conflated with meaning: if one is passed a note that reads 'Beth's grandmother is ill', written by someone who chose that sentence using a random device, then that sentence is meaningful, but carries no information about the state of health of Beth's grandmother. Even if the sentence is true, one cannot learn anything about Beth's grandmother from it.

Nevertheless, there's a tradition taking information as semantic, to the extent that (1) it eliminates possibilities, just as the truth of a meaningful (contingent) sentence is compatible with some possibilities and not others; and (2) it is about something, just as a meaningful sentence has a subject matter it addresses. These aspects have long been recognized, though emphasized in distinct traditions: compare *information-as-range* and *information-as-correlation* in Van Benthem and Martinez (2008).

An information source divides logical space into a partition of possibilities, and selects between them (definitively, if it is noise-free). What the information licenses as true is captured by the selection. What the information is about is captured by the distinctions that mark the borders of the partition. When the information source, a voice on the telephone, reports on the health of Beth's grandmother, Mary, it divides logical space, roughly, into cells such as: Mary is fit and hearty; Mary is under the weather; Mary has been hospitalized. It need not discriminate between issues such as: is Mary the grandmother of Sue, or is it Jane? Nor need it carry the information that

$2 + 2 = 4$, despite this being strictly implied by any true claim. Nor need it carry information about the source itself: it needn't report that the telephone connection is noise-free.

Whether semantic information is factive is the subject of a lively debate (Floridi 2019). Dretske takes information to be perforce truthful: '*false* information and *mis*-information are not kinds of information – any more than decoy ducks and rubber ducks are kinds of ducks' (Dretske 1999, 45). The KRI semantics makes no such assumption, witness the invalidity:

$$K^\varphi \psi \nvDash \varphi$$

No KRI validity depends on the assumption that information must be true, and no KRI invalidity depends on the existence of false information. '$K^\varphi \psi$', recall, has a subjunctive conditional flavour: 'One would be in a position to know ψ, if the total (empirical) information were φ'. This may be true, intuitively, because receiving φ positions one to know ψ at all (nearby) worlds where φ is true.

Knowledge must, of course, entail truth, even when information can fail to. The characteristic KRI condition (C0), $|\varphi| \subseteq f_\varphi(w)$, delivers the factivity of knowledge in the form:

(Factivity) $\{K^\varphi \psi, \varphi\} \vDash \psi^2$

When one is in a position to know that ψ given the information that φ, and φ *is* true, ψ must be true as well.

4.3   Unwise Adjunction

Accounts of knowledge that fail full closure under entailment or logical consequence face the issue of 'egregious violations': implausible failures of specific kinds of closure, unwarranted by the reasons that seemed to motivate rejecting full closure to begin with – fallibilism, escape from Cartesian scepticism, etc.: see Holliday (2012) for a thorough discussion. Kripke (2011a) finds egregious failures in the tracking notion of

---

[2]*Proof*: let $w \Vdash \varphi$ and $w \Vdash K^\varphi \psi$. By the former, $w \in |\varphi|$ so (C0) applies: $w \in f_\varphi(w)$. Then by the latter and (SK), $w \Vdash \psi$.

knowledge due to Nozick (1981), because it does not deliver
that one who knows a conjunction be positioned to know the
conjuncts. KRI is free from such a defect. Like all TSIMs, it
satisfies Simplification (from $K^\varphi(\psi \wedge \chi)$ to $K^\varphi\psi$ and $K^\varphi\chi$).
Like all TSIMs, KRI also satisfies Adjunction, i.e., the
entailment from $\{K^\varphi\psi, K^\varphi\chi\}$ to $K^\varphi(\psi \wedge \chi)$. It has been
argued that Adjunction is generally well-motivated for our
TSIMs, insofar as they stand for all-or-nothing attitudes, not
linked to the passing of an intermediate (say, probabilistic)
threshold (see Section 3.2). But what about the Makinson
(1965) Preface Paradox? In the original formulation, this
involves belief, and reasons to believe. The author of a (non-
fiction) thoroughly well-researched book justifiably believes
that each statement, $\varphi_1, ..., \varphi_n$, made in any proper chapter of
the book, is true. If Adjunction holds, the author is supposed
to believe that $\varphi_1 \wedge ... \wedge \varphi_n$. However, the author admits in the
preface, with proper epistemic modesty, that some of those
statements will be false (you know: 'Thanks to the anonymous
reviewers of the book... The remaining errors are mine'); and
so that it is not the case that $\varphi_1 \wedge ... \wedge \varphi_n$.

In spite of being adjunctive, the TSIM setting may accommodate, unlike standard epistemic logic, some intuition
in the ballpark of the Preface Paradox. A principle akin to
Adjunction is invalid in the TSIM semantics:

(Unwise Adjunction) $\{K^{\varphi_1}\psi, K^{\varphi_2}\chi\} \nvDash K^{\varphi_1 \wedge \varphi_2}(\psi \wedge \chi)$

Here's a second take on the Paradox in terms of KRI:
take any preface case where the agent is positioned to know
the claims, i.e., the conjuncts. As the agent is assumed to
reasonably believe, on inductive grounds, that there is likely
an error somewhere in the book, they are reasonable to accept
the negation of the big conjunction, and so, presumably,
are not positioned to know the big conjunction. KRI can
accommodate this with Unwise Adjunction. The proffered
explanation is that, if the informational bases for knowing
the conjuncts are suitably diverse, then the conjunction
of such bases might not be sufficient for knowing the big
conjunction, perhaps on the grounds that knowing a very
complex claim, involving highly complex subject matter, requires extra/special informational resources (e.g., to establish

higher-order epistemic facts, such as one's cognitive resources being sufficient for the extra book-keeping required).

4.4   The Stability of Knowledge

The KRI TSIM fails Monotonicity, the entailment from $K^{\varphi}\psi$ to $K^{\varphi \wedge \chi}\psi$. We've seen above how this handles one half of the dogmatism paradox. Now here's a toy example drawing on Hawthorne (2004), 71.[3] Assume that the information at issue is veridical. At the actual world, say @, one reads in *The Times* that Manchester United won. Now use M for the proposition that Manchester United won and T for the proposition that *The Times* reported that M. *The Times* is a trusted and reliable source, that offers a correct report. Hence, one is informed that $M \wedge T$ and thereby comes to know that $M \wedge T$. We can model this with a set-selection function: $f_{M \wedge T}(@) = |M| \cap |T|$, with $@ \in |M| \cap |T|$. Hence: $@ \Vdash K^{M \wedge T}M$.

But then one reads *The Globe*, which reports that Manchester United lost. This, unbeknownst to them, is a rare instance of a misprint in *The Globe*, which is itself trusted and reliable. Hence, *The Globe* is uninformative about the game's outcome, i.e., on the question of M. Nevertheless, glancing at the report yields some new information for one, G: *The Globe* reported a loss.

It seems that receiving this new information undermines one's knowledge that M. One should rationally suspend judgement on this claim. This is modelled as follows: $f_{M \wedge T \wedge G}(@) = |T| \cap |G|$, with $@ \in |T| \cap |G|$ and $|M| \cap |T| \cap |G| \subsetneq |T| \cap |G|$. Note that this accords with our condition (C0), since $|M| \cap |T| \cap |G| \subseteq f_{M \wedge T \wedge G}(@)$: the information $M \wedge T \wedge G$ leaves

---

[3]Hawthorne's example is similar, but developed with a different purpose: to serve as a puzzle about closure. He acknowledges that the puzzle is essentially the closure-based sub-paradox of the dogmatism paradox. Hawthorne's verdict is that closure can be preserved: knowing that φ puts one in a position to know that any evidence against φ is misleading, but the latter is 'junk knowledge' that is destroyed if new evidence is actually received, as in Sorensen (1988) (see Sharon and Spectre (2010) for a push-back). This last part indicates that Hawthorne (2004) advocates a rejection of Monotonicity.

only $T \wedge G$-worlds epistemically accessible, but allows for some $\neg M$-worlds. Hence: @ $\Vdash K^{M \wedge T \wedge G} G$ but @ $\Vdash \neg K^{M \wedge T \wedge G} M$.

If false information is allowed, the situation is described differently: since both *The Globe* and *The Times* are reliable and trusted, they both furnish information on the question of M. However, they conflict, yielding M and $\neg M$, respectively. The total information is thus $T \wedge G \wedge M \wedge \neg M$. Presumably, knowledge of M cannot be achieved here: the conflicting pieces of information cancel each other out. Hence: @ $\Vdash K^{M \wedge T} M$ but @ $\Vdash \neg K^{T \wedge G \wedge M \wedge \neg M} M$. This is modelled with $f_{T \wedge G \wedge M \wedge \neg M}(@) = |T| \cap |G|$.

Fallibilist epistemologists who claim that knowledge can be lost, e.g., due to the addition of defeaters (Brown 2018), face the challenge of accounting for the stability of knowledge: we want some way of preserving the aforementioned venerable Platonic insight that, unlike shaky opinion, *epistéme* cannot be overturned. KRI preserves it thanks to its characteristic condition (C0), which gives us another validity on top of the basic TSIM setting:

(Transitivity) $\{K^\varphi \psi, K^\psi \chi\} \vDash K^\varphi \chi$[4]

The case for Monotonicity was that it captures the core idea of stability. KRI suggests a different hypothesis: knowledge is stable in that it respects Transitivity. Suppose χ is known on the basis of information ψ, and that one's information is refined insofar as new information φ is received upon which knowledge of ψ can be based. Transitivity says that χ is still knowable: no knowledge is lost in the update from ψ to φ. This echoes (Dretske 1999, 57)'s Xerox Principle: if φ carries the information that ψ, and ψ carries the information that χ, then φ carries the information that χ.

Now take a version of the dogmatism paradox that hinges on Transitivity. With P, E, R, M as in Section 4.1:

$$K^{P \wedge R}(E \supset M) \text{ and } K^{E \wedge P \wedge R}(P \wedge R)$$

[4]*Proof:* assume that $w \Vdash K^\varphi \psi$ and $w \Vdash K^\psi \chi$. Thus: $\forall w_1 (w R_\varphi w_1 \Rightarrow w_1 \Vdash \psi)$ & $t(\psi) \leq t(\varphi)$ and $\forall w_2 (w R_\psi w_2 \Rightarrow w_2 \Vdash \chi)$ & $t(\chi) \leq t(\psi)$. Then $t(\chi) \leq t(\psi) \leq t(\varphi)$. By (C0), $|\psi| \subseteq f_\psi(w)$ and, by (SX), $f_\psi(w) \subseteq |\chi|$. Thus, $|\psi| \subseteq |\chi|$. By (SX) again, $f_\varphi(w) \subseteq |\psi|$. Hence, $f_\varphi(w) \subseteq |\chi|$.

That is: suppose that the joint information that P is true and that E supports $\neg P$ renders it knowable that E is misleading if true; and that refining the information to $E \wedge P \wedge R$ renders it jointly knowable that P and that E supports $\neg P$. Transitivity delivers that an agent with the refined information is positioned to know that E is misleading if true:

$$K^{E \wedge P \wedge R}(E \supset M)$$

When generalized, this seems objectionable. However, defiance in the style of Harman (1973) is here best interpreted as doubt about the truth of $K^{E \wedge P \wedge R}(P \wedge R)$. That an agent has received, in total, the information that $E \wedge P \wedge R$ need not position the agent to know P: the resultant knowledge that E is true and E supports $\neg P$ defeats rational belief in P. Defiance in the style of Sharon and Spectre (2010) is here best interpreted as doubt about the truth of $K^{P \wedge R}(E \supset M)$. That an agent has received, in total, the information that $P \wedge R$ cannot, in general, position the agent to know that E is misleading if true. Thus, standard responses to the paradox provide little motivation for rejecting Transitivity.

An advocate of inductive knowledge might be suspicious of Transitivity (we owe this to Alexandru Baltag). Let S be the (true) claim that Smoke is rising above the treeline, along with background information on the frequent correlation between smoke and wildfire. Let F be the true claim that there is a raging Fire in the forest. Let C be the claim that there is an inhabited cabin in the vicinity, with a Chimney leading from its fireplace. S, we suppose, provides inductive knowledge of F, in the absence of defeaters. Further, we suppose that C is exactly such a defeater. Hence an alleged counter-example to Transitivity

$$K^{S \wedge C}S \text{ and } K^{S}F, \text{ but } \neg K^{S \wedge C}F$$

To receive the information that there is smoke positions one to know there is (smoke and) fire, unless defeating information is also received.

We reject this counter-example: the above formalization seems a poor representation of the scenario at issue. That smoke signals fire is analogous to a voice on a telephone signalling that Beth's grandmother is ill, the headline of *The Times* signalling that Manchester United won, or Koplik spots signalling that a patient has measles. The former situation carries information about the latter. Coming to know that there is fire on the basis of smoke is like coming to know grandma is ill from a telephone call: the information that F is thereby transmitted, in a manner conducive to knowledge. To subsequently learn of the cabin is to lose knowledge of F despite having received the information that F, just as one loses the knowledge that grandma is ill when given a reason to doubt the testimony of the speaker, or doubt the quality of the telephone line.[5]

Such thinking is central in philosophical theories of information: the idea that information about a situation may flow to a receiver via a second situation – a *carrier* – is prominent in (Dretske 1999, ch. 5), (Skyrms 2010, ch. 3) and situation theory (Barwise and Etchemendy 1987; Barwise and Seligman 1995; Van Benthem and Martinez 2008; Seligman 2014). Consider:

> At this point some philosophers will say 'You might as well say that Smoke carries information about fire'. Well, doesn't it? Don't fossils carry information about past life forms? Doesn't the cosmic background radiation carry information about the early stages of the universe? *The world is full of information.* (Skyrms 2010, 44)

A better formalization of the above scenario does not bear on Transitivity

$$K^{S \wedge F \wedge C} S \text{ and } K^{S \wedge F} F, \text{ but } \neg K^{S \wedge F \wedge C} F$$

To receive the information that there is smoke is to receive the information that there is fire, positioning one to know there is (smoke and) fire, unless defeating information is also received. Finally, if sceptical that smoke carries the information that

---

[5] Here evidence and information may seem to pull apart. F, let's say, becomes part of one's information when one sees and correctly interprets the smoke. However, F does not seem to be part of one's evidence: rather, knowledge that F seems inferentially based one's evidence, e.g., the appearance as of smoke.

there is a wildfire for agents that know of the cabin, one might prefer:

$$K^{S \wedge C} S \text{ and } K^{S \wedge F} F, \text{ but } \neg K^{S \wedge C} F.$$

4.5   Closure Under (Known) Implication

Like all TSIMs, KRI fails Closure under Strict Implication (from $K^{\varphi}\psi$ and $\psi \prec \chi$ to $K^{\varphi}\chi$). We've seen above how this handles the other half of the dogmatism paradox. The issue with strict implication, as we know since Section 3.4 of the previous chapter, is that it can fail topic-preservation. In the KRI setting this is, in particular, preservation of the issues one's information allows one to address: although all the ψ-worlds are χ-worlds, thus all the φ-selected ψ-worlds are χ-worlds, φ may be information about the topic of ψ yet not be information about the topic of χ. Then, given φ, one can come to know ψ but not χ even if there just is no way for ψ to be true while χ is not.

Failures of such Closure are notoriously friendly to faillibilist epistemologies attempting to cordon off Cartesian scepticism. Take Dretske (1970) again: one's ordinary empirical information, delivered via sensory perception, puts one in the position to know one has hands. Having hands is incompatible with the possibility that one is a bodiless brain in a vat whose phenomenal experience is systematically misleading. Yet it might seem implausible that ordinary empirical information puts one in a position to rule out brain-in-vat scenarios.

Supporting fallibilist insights was already anticipated in Section 3.3 as one motivation for failing Addition (from $K^{\varphi}\psi$ to $K^{\varphi}(\psi \vee \chi)$), even for perfect reasoners to whom, additionally, the topic of χ is not alien. Let us develop this further. We know since chapter 2 that topics or subject matters are closely associated with sets of distinctions, issues, or questions (Lewis 1988a; Yablo 2014; Hawke 2018). To say that KRI is topic-sensitive is just to say that what one would be in a position to know if one were given a certain body of information depends on what that information can be about: what distinctions it speaks to; what issues it resolves, or, leaves open.

Now, the most compelling counter-examples to full epistemic closure can be understood as counter-examples to Addition, rooted in an enrichment of topic or subject matter. Recall Section 3.3: $\varphi \vee \neg\psi$ is equivalent to $\neg(\neg\varphi \wedge \psi)$ twice over, that is, both truth-conditionally and *qua* topic. Then the validity of Addition would commit one to:

$$K^\varphi\psi \vDash K^\varphi\neg(\neg\psi \wedge \chi)$$

But various cases impress philosophers as counter-examples to this principle – at least those who resist radical scepticism or Moorean dogmatism.[6] Knowing that one has hands, based on one's information, does not put one in a position to know that one is not a handless envatted brain (Cohen 1988). Knowing that the wall before one is red based on the visual information of it looking red does not put one in a position to deny that the wall is not red but subject to trick lighting (Cohen 2002). Knowing that the animal in the zebra enclosure is a zebra, based on the visual information that it looks like a zebra, does not put one in a position to know that the animal is not a cleverly disguised non-zebra (Dretske 1970, 2005). Or, back to our previous example: knowing that Kripke produced a result in modal logic, based on testimony in the classroom, does not put one in a position to deny the veridicality of a letter signed by Kripke claiming that he is a fraud.

We already mentioned how deniers of full epistemic closure, while in the business of cordoning off Cartesian scepticism, need to cordon off 'egregious failures' as well. Now KRI seems to invalidate the right things: Simplification tells us that $K^\varphi(\psi \wedge \chi)$ ensures $K^\varphi\psi$, but the failure of Addition tells us that $K^\varphi\psi$ does not ensure $K^\varphi(\psi \vee \chi)$. We motivated the asymmetry in the general TSIM set-up of the previous chapter. In the KRI setting: while the former appears indisputable, it is far from clear that knowability is closed under the introduction of arbitrary disjuncts: intuitively, the received information may not be about the topic of the alien disjunct.

---

[6] For further discussion, see Hawke (2016). For an opposing verdict, see Roush (2010) for a nuanced defence of the validity of the above principle. For push-back, see Avnur et al. (2011) and Hawke (2017, sect. 3.4.5).

Or suppose that one rejects unrestricted closure on the basis that various epistemic paradoxes (the dogmatism paradox, some evil demon Cartesian paradox) are best interpreted as counter-instances. One should then hope to invalidate any instance of Closure under $\prec$ that can be used to construct such paradoxes. Now the KRI semantics provides the following (easily, via failure of topic-preservation):

$$\{K^\varphi\psi, \psi \prec \chi\} \nvDash K^\varphi(\psi \wedge \chi)$$

If this principle (which was first given the spotlight, as far as we know, in Hawke (2016), sect. 4.2) were valid, then various paradoxes could be constructed. Suppose that $K^\varphi(P \wedge R)$ and $(P \wedge R) \prec (E \supset M)$, where P, E, R, M are as in the Kripke-Harman dogmatism paradox in Section 4.1 above. Then we could conclude: $K^\varphi((P \wedge R) \wedge (E \supset M))$. In other words, if it is known both that P and that E is generally a reason to reject P, then we could draw the dogmatic conclusion that it is knowable that P, that E is generally a reason to reject P and that if E were true then E would be misleading evidence. This dogmatic conclusion seems no better than that in the original puzzle.

On the other hand, closure under *known* material implication does hold – and for good reasons. In the KRI setting, call this principle *Closure Over Known Implication and Topic*

$$\text{(COOKIT)} \ \{K^\varphi\psi, K^\varphi(\psi \supset \chi)\} \vDash K^\varphi\chi \ ^7$$

COOKIT should hold. Here, both ψ and $\psi \supset \chi$ are fully on-topic with respect to the information that φ. Relative to that information, one is in a position to know both that ψ and that if ψ is true, χ is. Then one is in a position to know that χ, relative to the *same* information φ. (Given that KRI *qua* TSIM is non-monotonic, the inference may fail if the index for the available information is allowed to change across the involved formulas.) If, for instance, your information puts you

---

[7]*Proof*: let $w \Vdash K^\varphi\psi$ and $w \Vdash K^\varphi(\psi \supset \chi)$. By the former and (SX), for all w_1 such that $wR_\varphi w_1$, $w_1 \Vdash \psi$, and $t(\psi) \leq t(\varphi)$. By the latter and (SX) again, for all w_1 such that $wR_\varphi w_1$, $w_1 \Vdash \psi \supset \chi$. Thus for all w_1 such that $wR_\varphi w_1$, $w_1 \Vdash \psi$. Also, $t(\psi \supset \chi) = t(\psi) \oplus t(\chi) \leq t(\varphi)$, thus $t(\chi) \leq t(\varphi)$. Thus by (SX), $w \Vdash K_\varphi\chi$.

in the position to know both that Peano's postulates are true and that if these are then Goldbach's conjecture is, then you will also be in a position to know Goldbach's conjecture.

Authoritative sympathizers of full knowledge closure stress the plausible claim that the conclusion of a valid deductive argument from premises that remain known throughout must result in knowledge: see Williamson (2000), 118; Hawthorne (2004), sect. 1.5; Kripke (2011a), 200. There can't be such a thing as committing the fallacy of logical deduction, Kripke insists. This is the basis for the entire enterprise of mathematics: few want to deny the epistemic sanctity of mathematical results. This is often translated into a conviction in Closure under $\prec$, at least for logically astute agents. For, the rationale goes, the truth of $\psi \prec \chi$ is best understood in the setting of epistemic logic as an *a priori* truth of some kind.

Now a proponent of KRI need not deny the claim that deduction is a sanctified means for extending knowledge. She can dispute, however, that Closure under $\prec$ best captures this, given apparent counter-examples that can be extracted from epistemic paradoxes. Instead, she posits COOKIT as the uncontroversial core of the intuition.

How does acceptance of COOKIT not court trouble with regards to epistemic paradoxes? Take our case study again: Kripke-Harman dogmatism. The story is told as follows: suppose one has the information that $P \wedge R$ at time t_0. Further, since it is knowable *a priori* that $(P \wedge R) \prec (E \supset M)$, it is also knowable *a priori* that $(P \wedge R) \supset (E \supset M)$, and hence knowable on the basis of $P \wedge R$ that $(P \wedge R) \supset (E \supset M)$. But then COOKIT yields that one is in a position to know, on the basis of $P \wedge R$, that E must be misleading if true.

This reasoning betrays a confusion. A proponent of KRI need not accept that if it is knowable *a priori* that $(P \wedge R) \supset (E \supset M)$, then it is knowable on the basis of $P \wedge R$ that $(P \wedge R) \supset (E \supset M)$. That's because she need not accept that if φ is knowable *a priori* then φ is knowable on the basis of every body of information ψ. This is not licensed by the reading of 'on the basis of' that has been exploited. It is knowable *a priori* that $2 + 2 = 4$. It would be odd to conclude that $2 + 2 = 4$ can be known on the basis of the news that Beth's grandmother is ill.

The KRI setting can embed an absolute notion of a priority: what can be known without any empirical information by a computationally unbounded, logically astute agent with the full repertoire of concepts. Contrast a relative notion of a priority: what can be known without empirical information given a fixed, possibly incomplete, universe of concepts. Let $\top$ denote one's favourite tautology and let us read '$\top \supset \varphi$' as 'φ is *a priori*': φ is knowable *a priori* exactly when conceptual limitations are forgotten and φ is true at every possible world (let us flag that this requires worlds to be understood as representing epistemic possibilities, possibly in contrast to absolute or metaphysical possibilities).

With this in mind, a proponent of KRI can judge the case from the dogmatism paradox as follows. First, the case is best described as:

$$K^{P \wedge R}(P \wedge R)$$

$$\neg K^{P \wedge R}((P \wedge R) \prec (E \supset M))$$

$$\top \prec ((P \wedge R) \prec (E \supset M))$$

$$\neg K^{P \wedge R}(E \supset M)$$

Although $P \wedge R$ is knowable on the basis of the empirical information $P \wedge R$, it is not knowable on this basis that $(P \wedge R) \prec (E \supset M)$. Rather, this is *a priori*. In particular, this knowledge is based on concepts that go beyond those that comprise the topic of $P \wedge R$. In this case, in accord with COOKIT, the KRI subscriber can deny that one is positioned by one's empirical information to know that $E \supset M$.

4.6   Minimal Conditional Logic

The TSIMs introduced in the previous chapter, we know, are conditional-like and non-monotonic. From the viewpoint of conditional and non-monotonic logic, however, KRI is a peculiar operator. We close the chapter by discussing three principles KRI does not validate. Gabbay (1985) proposes them as a minimal foundation for a logic of non-monotonic derivations. In particular, they hold appeal as a base logic of *ceteris paribus* conditionals. They matter also because other TSIMs we'll meet in later chapters *will* validate them. We need to explain why, in the KRI reading, they rightly fail:

(Reflexivity) $\not\models K^\varphi \varphi$

(Cautious Transitivity) $\{K^\varphi \psi, K^{\varphi \wedge \psi} \chi\} \not\models K^\varphi \chi$

(Cautious Monotonicity) $\{K^\varphi \psi, K^\varphi \chi\} \not\models K^{\varphi \wedge \psi} \chi$[8]

As for Reflexivity, take the following line of reasoning:

1. $K^\psi \chi$ (Assumption)

2. $K^{\varphi \wedge \psi} (\varphi \wedge \psi)$ (by Reflexivity)

3. $K^{\varphi \wedge \psi} \psi$ (by Simplification)

4. $K^{\psi \wedge \psi} \chi$ (by Transitivity)

5. If $K^\psi \chi$ then $K^{\varphi \wedge \psi} \chi$ (by discarding 1)

If Reflexivity is conjoined with background principles we found to be independently good for KRI, we validate Monotonicity, exactly in the form the Kripke-Harman dogmatism paradox calls into question. One who accepts Reflexivity must either reject a Harman-like response to the dogmatism paradox, or bear the cost of rejecting Simplification, or Transitivity.

Besides, Reflexivity has direct counterexamples. This is obvious if information can fail to be factive: if an agent's total information I has a false part, then factivity assures that the agent does not know that I. But there are plausible counterexamples even when restricting to veridical information: examples linked to the failure of Monotonicity can be adapted to this effect.

---

[8] *Countermodels*: take atoms p, q (in all of the following, topic-assignments don't matter). First, Reflexivity. Let $W = \{w_1, w_2\}$, let $|p| = \{w_1\}$ and let $f_p(w_1) = W$. It follows that $w_1 \not\Vdash K^p p$. Second, Cautious Monotonicity. Let $W = \{w_1, w_2\}$. Let $|p| = W$ and $|q| = \{w_1\}$. Let $f_p(w_1) = |p|$ and $f_{p \wedge p}(w_1) = |q|$. It follows that $w_1 \Vdash K^p p \wedge K^{p \wedge p} q \wedge \neg K^p q$. Third, Cautious Transitivity. Let $W = \{w_1, w_2\}$. Let $|p| = \{w_1\}$. Let $f_p(w_1) = |p|$ and $f_{p \wedge p}(w_1) = W$. It follows that $w_1 \Vdash K^p p \wedge \neg K^{p \wedge p} p$.

Here is another: suppose Mary watches Joe Biden deliver his state of the union address, from a front row seat, hearing distinctly that his first topic is trade. A week later, Mary's memory of the speech remains vivid. Presumably, her senses informed her that his first topic was trade, she thereby came to know it, and she now preserves this knowledge via memory. However, an epistemic peer then claims that Biden's first topic was gun control, reminding Mary that her memory can be unreliable. Given this, it can be rational for Mary to suspend or weaken her belief that the first topic was trade, losing her knowledge. Nevertheless, it remains true, in an important sense, that Mary has the information that the first topic is trade (T): she received that information through a perceptual event that, at the time, was conducive to knowledge. The event and its interpretation remain vividly stored in her memory. So, if $I \wedge T$ is Mary's total information: $\neg K^{I \wedge T}(I \wedge T)$.

The KRI semantics invalidates Cautious Transitivity and Cautious Monotonicity because our TSIM set-selection functions are indexed to *formulas* and few constraints regulate how sets are selected for different formulas. Condition (C0) requires the sets selected for p and for $p \wedge p$, for instance, to both contain every p-world, but otherwise, no constraint is imposed. Models are allowed where, for instance, $|p| \subsetneq W$, $f_p(w) = |p|$ and $f_{p \wedge p}(w) = W$. (Things will be different in later chapters, when imposing other constraints on f_φ will secure that the φ input in $X^\varphi \psi$ actually boils down to $|\varphi|$, the thin proposition or set of worlds where φ is true.) Thus, the question as to whether Cautious Transitivity and Cautious Monotonicity should be treated as logical truths is bound up with substantive issues: does a piece of information have a logical structure, and in particular one that mirrors the syntax of a sentence with which it is expressed? If so, to what extent should an epistemic logic accommodate agents whose cognition is sensitive to syntax? One might wish to model agents whose capacity to extract knowledge from information tracks the complexity of the information's structure. This impulse is waged against an insistence that $f_{\varphi \wedge \psi}$ always selects the same set as f_φ when $|\varphi| = |\varphi \wedge \psi|$.

One possible view has it that information is unstructured. Or one might accept that information is structured, but hold that this structure should be ignored when dealing with KRI-idealized agents. Then since p and $p \wedge p$ have the same topic and truth set, they should be treated as equivalent. With this in mind, consider the class of models that satisfy a Twice Over Equivalence principle:

(TOE) $|\varphi| = |\psi|$ & $t(\varphi) = t(\psi)$ $\Rightarrow$ $f_\varphi(w) = f_\psi(w)$ for all w

TOE says that if φ and ψ are both co-necessary and about the same same topic, then their set-selection functions always output the same values. Models complying with TOE filter out a number of syntactic differences concerning the way information is presented, but still allow, via differences in topicality, hyperintensional distinctions involving pieces of information with coincident truth sets. It is easy to check that Cautious Monotonicity and Cautious Transitivity are validated if we impose this restriction on the admissible models. It may also be confirmed that, however, Monotonicity, in full generality, is *not* validated by this restricted class. Compliance with TOE allows that if

$$|\varphi \wedge \psi| \subsetneq |\varphi \wedge (\psi \vee \neg\psi)|$$

then the set selected for $\varphi \wedge \psi$ need not be a subset of that selected for $\varphi \wedge (\psi \vee \neg\psi)$, despite these sentences sharing a topic and the former entailing the latter.

4.7   Chapter Summary

This chapter has explored an epistemic reading of the two-place TSIM operators, in which they are interpreted as capturing the notion of KRI, $K^\varphi \psi$: one would be in a position to know ψ, if the total (empirical) information were φ. In the KRI setting, the variable strictness of two-place TSIMs makes them suitable to model the non-monotonicity of knowledge acquisition while allowing knowledge to be intrinsically stable. Their topic-sensitivity allows them to invalidate controversial forms of epistemic closure while validating less controversial ones. Unlike the standard Hintikkan modal framework for epistemic logic, the KRI framework models insights friendly to epistemic fallibilism, as well as accommodating plausible approaches to the Kripke-Harman dogmatism paradox which bear on non-monotonicity or on topic-sensitivity. We have described how, in the KRI reading, the TSIMs don't comply with certain requirements for non-monotonic conditionals, and we have argued that such non-compliance is justified.

5

Imagination and Suppositional Thought

Constraint (C1) from Section 3.5, $f_\varphi(w) \subseteq |\varphi|$, is the converse of KRI's (C0): (C1) says that all the φ-selected worlds are φ worlds, but allows for φ-worlds which are not selected. Read it as saying that any situation one looks at, given supposition φ, is one where φ holds. Read the relevant $wR_\varphi w_1$ as saying that w_1 is one of the worlds where things are as imagined (at w), given supposition φ. We then relabel our generic TSIM '$X^\varphi \psi$' as '$I^\varphi \psi$', for we take it as expressing imagination of a certain sort: we read it as 'Supposing that φ, one imagines ψ'; or, less tersely, as: 'In an act of imagination starting with suppositional input φ, one imagines that ψ'.

What kind of imagination is this? 'Imagining' is highly ambiguous: we use it for such mental activities as entertaining some idea, daydreaming, hallucinating, free mental wandering. Call the kind we are after *Reality-Oriented Mental Simulation* (ROMS; why 'reality-oriented', and why 'simulation', we will see soon). It is closely tied to suppositional thinking: we suppose that something is the case; develop the supposition importing background knowledge and beliefs; and imagine what things are like in the unpacked scenario.

Section 5.1 deals with a puzzle concerning how imagination, so understood, can have some cognitive value in spite of its apparent arbitrariness. It suggests that the puzzle can

Topics of Thought: The Logic of Knowledge, Belief, Imagination.
Francesco Berto, Oxford University Press. © Francesco Berto 2022.
DOI: 10.1093/oso/9780192857491.003.0005

be addressed by exploring the connection between ROMS and conditional beliefs. Section 5.2 lists a series of plausible features of ROMS based on literature in cognitive psychology and the philosophy of mind. Section 5.3 shows that the patterns of validities and invalidities involving our two-place TSIMs, in ROMS clothing, model those features.

Section 5.4 discusses the opportunity of adding condition (C2) from Section 3.5. The addition validates a restricted equivalence principle for TSIMs, which strengthens the logic and limits the hyperintensional anarchy of imagination: when φ and ψ are equivalent in imagination, one will imagine the same things after supposing either. Equivalence in imagination is cognitive equivalence: φ and ψ are cognitively equivalent for one when they play the same role in one's cognitive life – whatever one understands, concludes, etc., given either, one does, given the other. However, (C2) also validates the Cautious Transitivity principle from Section 4.6, whose desirability for a logic of ROMS, as we will see, might perhaps be found dubious.

Finally, Section 5.5 closes by discussing the opportunity of weakening the strict topic-inclusion requirement of our TSIMs, at least (and perhaps not only) when they present themselves in ROMS clothing.

5.1   Belief and the Anarchy of Imagination

In their introduction to the beautiful collection *Knowledge Through Imagination*, Amy Kind and Peter Kung state what they call 'the puzzle of imaginative use' (Kind and Kung 2016, 1): imagination seems to be arbitrary escape from reality; how can it give knowledge of reality?

This looks like a real dilemma, for both of the supposedly incompatible features of imagination – arbitrariness, epistemic value – are well-established. Here's Hume on arbitrariness:

> Nothing, at first view, is more unbounded than the thought of man, which not only escapes all human power and authority, but is not even restrained within the limits of nature and reality.

> To form monsters, and join incongruous shapes and appearances, costs the imagination no more trouble than to conceive the most natural and familiar objects. And while the body is confined to one planet, along which it creeps with pain and difficulty; the thought can in an instant transport us into the most distant regions of the universe; or even beyond the universe, into the unbounded chaos, where nature is supposed to lie in total confusion. (*Enquiry*, 2)

Yet Hume also thought, famously, that:

> 'Tis an establish'd maxim in metaphysics, that whatever the mind clearly conceives includes the idea of possible existence, or in other words, that nothing we imagine is absolutely impossible. (*Treatise*, I, ii, 2)

Thus, imagination is a pathway to knowledge of absolute modalities. I should mention that many, including Byrne (2007); Fiocco (2007); Jago (2014); Kung (2014); Priest (2016), think that, *contra* Hume, we can imagine the absolutely impossible. I agree: see Berto and Schoonen (2018). Of course, Hume's overall view is consistent: imagination is completely unbounded when it deals with recombinations of matters of fact, but bound by absolute necessity as captured by relations of ideas.

However, this stance won't quite solve the puzzle. It is widely agreed in cognitive psychology (Markman et al. 2009) that imagination is of epistemic value for more down-to-earth purposes than knowledge of absolute modalities. It allows us to improve performance; to make contingency plans by anticipating what will happen if such-and-so turns out to be the case ('What if I lose my job due to Brexit?'); to learn from mistakes by considering how we could have done better by acting differently, or to ascertain responsibilities ('Would he have been hit by the car had the driver respected the speed limit?'). It is invoked to account for counterfactual thought (Byrne 2005; Byrne and Girotto 2009) and pretence (Nichols and Stich 2003). In semantics and the philosophy of language, it is used to explain how we evaluate conditionals (Stalnaker 1968; Evans and Over 2004). In the philosophy of fiction, it can explain our engagement with fictional works (Walton 1990; Currie 1990). In epistemology, it is used to account for how we can come to know modal claims of various sorts (Yablo 1993; Chalmers 2002; Williamson 2007).

We *can* form reliable judgements around issues we explore in our imagination. But how can this be, if imagination is governed by free associations of ideas of the kind highlighted by Hume? An imagination operator will look like Prior's *tonk*: given input φ, it will output whatever ψ one likes. The motto 'Logic will take you from A to B, imagination will take you everywhere' sounds as if it says something good about imagination. But, well, at least logic will take you to the *proper B*s: those that follow logically from A. By taking you anywhere, imagination will take you to no place of added epistemic value.

The short version of the answer to the puzzle, shared by a number of philosophers in the Kind and Kung collection (Kind 2016; Langland-Hassan 2016; Williamson 2016a) is: imagination, of the kind that can give us knowledge, is *constrained*. It deviates from reality, but in a regimented way – hence the 'reality-orientation': What if Stauffenberg had put the bomb on the other side of the table? We suppose that he does but, when we're after a serious assessment of how things would have gone, we don't imagine that Hitler suddenly grows an armour protecting him from the detonation, or that the bomb becomes a vase of flowers. We follow some minimal alteration principle: we keep reality more or less as we know or believe it to be, compatibly with what's needed for our initial supposition to hold. This is sometimes called 'reality-monitoring' in cognitive psychology: see Johnson and Raye (1981): we 'use our knowledge of how the world works to help keep the imagining realistic' (Davies 2019, 187). An account of imagination as ROMS needs to represent both the deviation and the regimentation. To get them right, I submit, following Chris Badura (2021a), we need to explore the connection between ROMS and belief.

Various authors, e.g., Currie and Ravenscroft (2002), Goldman (2002), agree on a simulationist account of imagination:

imagination in general re-creates counterpart non-imaginative mental states. In perceptual imagination, one simulates perception: one hears the riff of *Sunshine of Your Love* in one's head (go on, try!) and it's relevantly like the real thing, except that one's auditive apparatus is not involved. Propositional imagination, the kind thereof which seems to have a chance, if any does, at helping with knowledge, is the one in which one simulates *belief*: Arcangeli (2019) calls it 'cognitive imagination'.

On the other hand, imagination is often contrasted with belief: the former is voluntary (Davies 2019, 92) in a way the latter isn't, at least according to most authors (Walton 1990; Mulligan 1999; Gendler 2000b; Dorsch 2012). One can imagine that all of Edinburgh has been painted yellow but, having overwhelming evidence of Edinburgh's greyness, one cannot make oneself believe it. What's the point of a simulation that differs from what it simulates in such a core feature?

The link between imagination as mental simulation, and belief, is, I think, that in mental simulation we imagine in order to assess what will happen *if* something is the case, or what would have happened *if* something had been the case. What if I jump the stream? Will I make it to the other side, or will I fall in the water and drown? The kind of belief mental simulation typically connects to for epistemic purposes, then, must be *conditional* belief.

I don't really believe, say, that Brexit will cause a recession. I explore the consequences, though: What if Brexit causes a recession? If I conclude, given the supposition that Brexit causes a recession and what I know or believe, that I will lose my job in the hypothetical scenario, I form the relevant conditional belief. This obviously relates to (what we now call) the Ramsey test for conditionals: one evaluates a conditional 'If φ, then ψ', by supposing the antecedent, φ, minimally adjusting one's belief or knowledge system, and seeing whether the consequent ψ turns out in the imagined scenario. Ramsey's mythical footnote:

> If two people are arguing 'If p will q' and are both in doubt as to p, they are adding p hypothetically to their stock of knowledge and arguing on that basis about q; so that in

a sense 'If p, q' and 'If p, $\neg q$' are contradictories. We can
say that they are fixing their degrees of belief in q given p.
(Ramsey 1990, 155n)

The Ramsey test ties our evaluation of conditionals to the
update of our prior beliefs in the light of new information.
It's around noon and one sees that John has moved into the
kitchen. One then minimally revises one's beliefs compatibly
with the news. In the updated belief system, one will have
dropped, say, the previously held belief that John is in the
living room; on the other hand, that John is cooking will now
look plausible enough for one to come to believe it. What
is typical of mental simulation, is that one doesn't really
get the news online, for instance, by seeing that John is in
the kitchen. One just pretends, in offline mode (Williamson
2016a), that John is there, and checks what would follow.
If in the hypothetical situation it turns out that John is
cooking, one does not believe that John is cooking; but one
can acquire a conditional belief: one believes that John is
cooking conditional on John being in the kitchen.

Does one come to believe the corresponding conditional,
too, 'If John is in the kitchen, then he's cooking'? After
Lewis (1976)'s well-known triviality results it is often agreed
that, however tight the connection between conditional belief and belief in the corresponding conditional, it won't
be identity. Philosophers and psychologists have tried to
circumvent the problem: some keep the core of the Ramsey
test by stating that conditionals don't express propositions
and cannot generally be embedded, e.g., Edgington (1995);
Bennett (2003). Psychologists often content themselves with
a robust, empirically corroborated correspondence, short of
identity, e.g., Evans and Over (2004); Oaksford and Chater
(2010). We will come back to these issues in chapter 8,
in the context of our theory of topic-sensitive indicative
conditionals.

Supposing φ is quite different from supposing that one
believes that φ. It's the first one that goes on, in general,
in mental simulation, although, of course, there will be
special cases where one supposes that one believes something.
Supposing I am prime minister of the UK, I imagine that I will
hire a bunch of friends in key governmental roles. Supposing I

believe I am prime minister of the UK, I imagine very different things, e.g., that I need to be hospitalized for I am seriously deluded. Supposing that my business partner is cheating on me, I may imagine that I'll never realize that he is. Supposing I believe that my business partner is cheating on me, I imagine that I have realized that he is – a variation on 'Thomason conditionals': see Van Fraassen (1980); Bennett (2003), 28-9.

Now a number of authors agree that not everything is arbitrary in imagination as mental simulation. The things that aren't, I think, can be tied to the non-arbitrariness of belief. In the following section, I try to tell which is which by listing a series of features that characterize imagination as ROMS, drawing on research in cognitive psychology and the philosophy of mind. In the sections after that, I show that the semantics of our TSIMs, in ROMS clothing, models all of them.

5.2   Reality-Oriented Mental Simulation

First: ROMS is episodic (Atance and O'Neill 2001) and agentive (Wansing 2017). In an imaginative episode, we start by supposing something, φ, carry out the exercise for a while, often by controlling some aspects of what we imagine (Van Leeuwen 2016), and stop after some time.

Second: such suppositional input is up to us. In their influential model of mental simulation, Nichols and Stich (2003) have 'an initial premise or set of premises, which are the basic assumptions about what is to be pretended' (24). This may be made up by the agent when engaging in predictions, e.g., when you guess what would happen if something were the case; or it may be taken on board via an external instruction, e.g., when you read a novel and take the explicit text as your input, or when you evaluate a conditional and start by taking the antecedent as input. In ROMS clothing, the two-place TSIMs represent the suppositional input as expressed by formulas in their first position.

Third: we integrate the input with background information we import, contextually, depending on what we know or believe (Davies 2019, 45-6). The importance of background knowledge and beliefs in suppositional thinking is increasingly acknowledged in the psychology of reasoning (Oaksford and Chater 2010). Once the input is in, Nichols and Stich (2003) claim, 'children and adults elaborate the pretend scenarios in ways that are not inferential at all', filling in the explicit instruction with 'an increasingly detailed description of what the world would be like if the initiating representation were true' (26-28).

When we imagine Watson talking with Holmes while walking through the streets of London, for instance, we are likely to represent Watson dressed as a nineteenth century gentleman, not as an astronaut. The text of the relevant novel need not say anything explicitly on how Watson is dressed, nor do we infer this from the explicit content via sheer deductive logic. Rather, we import such information into the represented situation, based on what we know or believe: we know that the story is set up as taking place in Victorian London and we assume, lacking information to the contrary from the text, that Watson is dressed as we believe gentlemen were dressed at the time. As Hannes Leitgeb has it, 'suppositional reasoning would be quite pointless if one were not able to supplement the assumed proposition by various background beliefs that are preserved by the act of assumption.' (Leitgeb 2017, 73).

Now there is some agreement on the idea that, whereas one can suppose whatever φ one wills (Arcangeli (2019) makes a forceful case), what ψs will come out true in the imagined scenario once the suppositional input is in is largely *not* up to us. It has to do with how our prior belief system is disposed to (minimally) adjust itself conditional on φ, and how this works is largely involuntary, as beliefs are. Thus, various authors think that the involuntary component of imagination as ROMS comes into play exactly here. As Williamson (2016a), 116, has it:

> Think of a hunter who finds his way obstructed by a mountain stream rushing between the rocks. He reaches the only place in the vicinity where jumping the stream might be feasible. [...] How should he try to determine whether he would succeed? [...] One *imagines* oneself trying. If one then imagines

oneself succeeding, one judges that if one tried, one would succeed. If instead one imagines oneself failing, one judges that if one tried, one would fail. [...] When the hunter makes himself imagine trying to jump the stream, his imagination operates in voluntary mode. But he neither makes himself imagine succeeding nor makes himself imagine failing. Rather, having forced the initial conditions, he lets the rest of the imaginative exercise unfold without further interference. For that remainder, his imagination operates in involuntary mode. He imagines the antecedent of the conditional voluntarily, the consequent involuntarily. Left to itself, the imagination develops the scenario in a reality-oriented way, by default.

Similarly, Langland-Hassan (2016) distinguishes between 'guiding chosen' imaginings, 'top down intentions [that] are key to initiating an imagining', and 'lateral constraints [that] govern how it then unfolds' (9), and which seem to operate in involuntary mode. If the additional details are borrowed from our knowledge or belief base, as Nichols and Stich (2003) and Van Leeuwen (2016) have it (the former have a cognitive 'belief box', from which contents are taken and imported into the mental simulation), this makes sense: for if beliefs are often formed and managed in largely involuntary mode, it seems plausible for their importation to be essentially involuntary. Some research in cognitive psychology seems to support the view that imagination allows automatic, involuntary access to the knowledge deposited in implicit (long-term) memory, and that the results of imaginative exercises can themselves alter such memory (Kosslyn and Moulton 2009).

Our TSIMs naturally represent this integration of the initial supposition via background beliefs and knowledge by being variably strict modals: the suppositional input φ has the imagining agent at w look at a bunch of worlds: those in $f_\varphi(w)$, where things are as one imagines them to be, given the input. We can think of such worlds as collectively representing how things would be for one, if the suppositional input were true.

Which worlds should we look at, exactly? What do the

items in $f_\varphi(w)$ have in common? The insight behind the minimal alteration principle would be that these are the most plausible worlds given one's belief or knowledge state. This, however, is not really represented in the TSIM semantics discussed below. To do it, one can impose a total (pre)ordering of worlds by plausibility, mimicking what is done in the Lewis (1973) sphere semantics and in various epistemic logics for belief revision – see e.g. Van Benthem (2007). This is exactly what I'll do in the next chapter 6, where I apply TSIM theory to model hyperintensional conditional belief and belief revision. One reason not to do it in the present context, is that a semantics with the plausibility ordering automatically satisfies constraint (C2) from Section 3.5. In Section 5.4, I want to discuss, instead, the opportunity of adding it manually.

Finally, fourth: imagination has topicality or relevance constraints. We do not indiscriminately import unrelated contents into the imagined scenarios: we focus on what's on-topic, given the input. As Amy Kind has it, '[We require] that the world be imagined as it is *in all relevant respects*' (Kind 2016, 153). This is key to distinguishing imagination as ROMS from free-floating mental wandering. You know that the US president is elected every four years, but this is immaterial to your imagining Watson and Holmes' adventures from Doyle's novels, insofar as they don't involve topics connected to the US presidency. So you will not import knowledge or beliefs in the ballpark, even when perfectly consistent with the suppositional input. When one supposes that Stauffenberg puts the bomb on the other side of the table, develops the supposition in imagination, and concludes that, then, Hitler gets killed, one doesn't imagine that either Hitler gets killed or there's life on Kepler-442b, although $\varphi \vee \psi$ is an immediate logical consequence of φ. One also doesn't imagine each irrelevant logical or necessary truth given that suppositional input, although such truths are implied, in classical logic, by anything, thus by any input, and such truths hold at all possible worlds, thus at all worlds where the input is true. This topic-sensitivity of ROMS is, of course, captured by the topic-inclusion filter of our TSIMs. (Isn't full topic *inclusion* too strong for imagination? Mustn't imagination, even of a regimented kind, be to some extent ampliative and creative? We'll get back to this at the end of the chapter.) Let us now explore the logic of ROMS.

5.3 The Logic of ROMS

Constraint (C1) has it that all the worlds accessible given φ will make φ true. This seems right for suppositional thinking: when one supposes that Stauffenberg puts the bomb on the other side of the table, one looks at a situation where Stauffenberg *does* put the bomb on the other side of the table to begin with. (C1) validates a reflexivity principle for our TSIMs:

(Success) $\vDash I^\varphi \varphi$

The name 'Success' comes from AGM belief revision theory – we'll talk about this in the next chapter. When one reads our ROMS TSIM as 'Given suppositional input φ, one imagines that ψ', Success may not be so straightforward due to possible cases of 'imaginative resistance' (Gendler 2000a): one may be prompted to imagine that φ by a work of fiction giving φ as an explicit suppositional input, but fail to imagine it. The debate on imaginative resistance is lively: see Gendler (2020) for a reconstruction. Some authors distinguish imagining and supposing, claiming, e.g., that imagining requires a 'commitment' which is lacking in supposition (Balcerak Jackson 2016). If so, there might be imaginative resistance even if, as I believe following Arcangeli, 'there is no such thing as "suppositional resistance"' (Arcangeli 2019, 52).

On the other hand, lacking (C0), the Factivity principle of Section 4.2, which held for KRI, fails for ROMS – and rightly so, for imagination, unlike knowability, isn't factive:

(Factivity) $\{I^\varphi \psi, \varphi\} \nvDash \psi$

Non-factivity is tied to false beliefs sneaking in. I suppose that Daniel is in South Bend, φ, and, as a matter of fact, he happens to be there. I develop the scenario in my imagination by importing my false background belief that South Bend is in Illinois. I imagine that Daniel is in Illinois. $I^\varphi \psi$. It doesnt follow that ψ is true, Daniel is in Illinois.

Some accounts of imagination give us further reasons for liking the fully conjunctive nature of TSIMs, validating the inferences from $I^\varphi(\psi \wedge \chi)$ to $I^\varphi \psi$, and from $I^\varphi \psi$ and $I^\varphi \chi$ to $I^\varphi(\psi \wedge \chi)$. This may sound particularly convincing if imagination essentially involves mental imagery, as famously argued by Kind (2001).

Unlike propositional or language-like mental representations, which are processed serially, the way we process sentences, pictorial mental representations have quasi-spatial features (Paivio 1986). Classic empirical work in psychology (Shephard and Metzler 1971; Pinker 1980; Block 1983), showed, e.g., that the time taken to scan between two points of a mental image is generally proportional to their subjective distance; that larger objects fill the imagined scenario sooner than smaller ones; etc. This may point at some mereological, quasi-spatial structure represented in the mind: when we visually imagine our bedroom, we can zoom into one part of the scene (where the bedside table is), then zoom into one further sub-part (the book on the bedside table), then move upwards, etc.

Maybe, then, the conjunctive nature of imagination is supported by such considerations on the intuitive quasi-spatio-mereological structure of mental imagery: maybe you can't imagine that the table is brown and square without imagining that it's brown because, when you pictorially imagine the whole situation, you imagine its parts; and you can't imagine that the table is brown and that the table is square without imagining (in one go, that is, once the relevant contextual parameters are fixed: see Section 3.2) that the table is brown and square, because when you pictorially imagine the parts all together, you imagine the whole. Think about physical picturing. To picture something *is* to picture its parts, and there is no more work to do then: once all the parts are pictured, the whole is, automatically.

It is, however, controversial (Van Leeuwen 2013; Williamson 2016a; Gregory 2016) that imagination essentially involves mental imagery, even if one admits that some mental representations represent pictorially, which is itself a controversial claim (see the so-called imagery debate, e.g., Pylyshyn (1981, 2002)). Cognitive psychologists distinguish purely 'conceptual imagination' (Davies 2019, 3) or 'suppositional imagination' from 'enactment imagination' involving mental imagery (Nichols 2006, 41-2). In mental simulation we sometimes imagine scenarios involving only abstract objects: think of a mathematician mentally going through a proof, without making use of quasi-spatial structures like graphs or so. Perhaps we imagine, on occasion, complex situations whose representation needn't involve mental imagery: think of yourself trying to predict how the markets worldwide will react to the next economic downturn. One may not want to rest one's case for Simplification and Adjunction just on considerations concerning the supposedly essentially imagistic nature of imagination.

Four invalidities holding generally for our two-place TSIMs make good sense for ROMS in particular. First, their variable strictness models the non-monotonicity of suppositional thinking. $I^\varphi\psi$ does not entail $I^{\varphi\wedge\chi}\psi$: supposing the weather is nice tomorrow, you imagine you will go play football. Supposing more, namely that the weather is nice and you have a hundred essays to mark, you don't imagine that.

Second, imagination, like intentional states in general, can under-determine its contents. $I^\varphi(\psi \vee \chi)$ does not entail that $I^\varphi\psi \vee I^\varphi\chi$: supposing Mary is in New Zealand, you imagine that she'll be either in the North Island or in the South Island, but you don't imagine that she is in the former in particular, nor do you imagine she's in particular in the latter.

Third, imagination isn't additive. And so $I^\varphi\psi$ does not entail $I^\varphi(\psi \vee \chi)$: when, starting from the supposition that Stauffenberg puts the bomb on the other side of the table, you imagine that Hitler gets killed, you don't thereby imagine that either Hitler gets killed or there's life on Kepler-442b – and rightly so, insofar as you don't want to clutter your mind with thoughts that are off-topic with respect to the goal of the suppositional exercise. Imagination is topic-sensitive, and topics, we know, are naturally connected to issues or questions under investigation: adding the Kepler-442b disjunct will not mark any progress with the question addressed in the suppositional exercise (plausibly: What happens to Hitler?).

Fourth, imagination is hyperintensional. In particular, it still fails the Equivalence principle from Section 3.4. Its logic is rather weak, but we can make it stronger by adding constraint (C2) from our list in Section 3.5, with the effect of limiting the hyperintensionality of ROMS. Whether we should add (C2) is the topic of the next section.

5.4   Equivalence in Imagination

Here's the idea: although it is not the case that, given two intensionally equivalent suppositional inputs, φ and ψ, one will imagine the same things, this will happen when φ and ψ are at least equivalent in one's imagination.

What is equivalence in imagination? We can understand it as cognitive equivalence or synonymy, which is different from synonymy *tout-court*: φ and ψ are cognitive equivalents or cognitive synonyms for one when they play the same role in one's cognitive life; I got the idea from Hornischer (2017). Roughly: whatever one concludes – deductively, abductively, inductively – supposing either, one does, supposing the other. Whatever one understands when either is uttered, one does, when the other is. One cannot take either as true, without taking the other as true. Whatever one thinks about given either, one does, given the other.

Cognitive synonymy seems to be a respectable idea in linguistics. Absolute synonymy, understood as substitutivity *salva veritate* in all contexts, has notoriously raised eyebrows among philosophers. Linguists sometimes take it as a purely theoretical, limit notion, for it cannot be empirically tested (Stanojević 2009; Cruse 2017). On the other hand, cognitive synonymy is the working concept for an amount of research in linguistic semantics (Lyons 1996; Murphy 2003).

Cognitive equivalence should be relative to the thinker's available knowledge-belief base and storage of concepts. *John is a bachelor* and *John is an unmarried man* are equivalent for nearly any competent speaker of English: if one takes either as true but the other as false, this is likely to generate suspicion on the level of English proficiency of the speaker. *Ex contradictione quodlibet is classically valid* and *Pseudo-Scotus' Law is classically valid* are equivalent for most logicians. *Groundhogs are rodent* and *Woodchucks are rodent*, for most zoologists. One way to represent such cognitive interchangeability in AI is via pairs of defeasible, non-monotonic conditionals, 'If φ, ψ', 'If ψ, φ', stored in the agent's knowledge base, e.g., in Logic Programming (Stenning and van Lambalgen 2008).

Now our target constraint, recall, goes thus:

(C2) $f_\varphi(w) \subseteq |\psi|$ & $f_\psi(w) \subseteq |\varphi| \Rightarrow f_\varphi(w) = f_\psi(w)$

If all the φ-selected worlds make ψ true and vice versa, then φ and ψ are equivalent in imagination: when one takes either as one's suppositional input in an act of mental simulation, one looks at the same circumstances.

(C1) and (C2) together make the suppositional input φ in $I^\varphi \psi$ way less syntactic: they entail that, for all w, if $|\varphi| = |\psi|$, then $f_\varphi(w) = f_\psi(w)$, that is, when φ and ψ are intensionally equivalent, their set-selection functions coincide.[1] This makes the input φ boil down to the corresponding thin proposition $|\varphi|$ (compare the discussion of the TOE principle in the KRI setting, in Section 4.6).

More importantly, (C2) validates the nice principle promised above, limiting the hyperintensional anarchy of imagination.

(Restricted Equivalence) $\{I^\varphi \psi, I^\psi \varphi, I^\varphi \chi\} \vDash I^\psi \chi$[2]

This says that equivalents in imagination φ and ψ can be replaced *salva veritate* as indexes in $I^{...}$. This seems right, and desirable in a logic of suppositional thought, in spite of the many hyperintensional distinctions we may draw in our imagination. For say that *bachelor* and *unmarried man* are for you equivalent in imagination: you are so firmly aware of their meaning the same, that you cannot imagine someone being either without imagining him being the other ($I^\varphi \psi$ and $I^\psi \varphi$ entail $t(\varphi) = t(\psi)$: equivalents in imagination are always

---

[1]Suppose $|\varphi| = |\psi|$; then by (C1), $f_\varphi(w) \subseteq |\varphi| = |\psi|$ and $f_\psi(w) \subseteq |\psi| = |\varphi|$; then, by (C2), $f_\varphi(w) = f_\psi(w)$: see (Priest 2008, 94).

[2]*Proof:* suppose $w \Vdash I^\varphi \psi$, $w \Vdash I^\psi \phi$, $w \Vdash I^\varphi \chi$. By (SX), these entail, respectively, (a) $f_\varphi(w) \subseteq |\psi|$ and $t(\psi) \leq t(\varphi)$, (b) $f_\psi(w) \subseteq |\varphi|$ and $t(\varphi) \leq t(\psi)$, (c) $f_\varphi(w) \subseteq |\chi|$ and $t(\chi) \leq t(\varphi)$. From (a) and (b) we get $f_\varphi(w) = f_\psi(w)$ (by (C2)) and $t(\varphi) = t(\psi)$ (by antisymmetry of topic parthood). From these and (c) we get $f_\psi(w) \subseteq |\chi|$ and $t(\chi) \leq t(\psi)$. Thus by (SX) again, $w \Vdash I^\psi \chi$.

about the same topic for the thinking subject). Thus, $I^\varphi \psi$, when you imagine that John is unmarried, you imagine that he is a bachelor, and $I^\psi \varphi$, when you imagine that John is a bachelor, you imagine that he is unmarried. Suppose $I^\varphi \chi$: as you imagine that John is unmarried, you imagine that he has no marriage allowance. Then the same happens as you imagine that he is a bachelor, $I^\psi \chi$. All works smoothly – so far.

In Section 3.5, however, I said that that the addition of (C2) in the current setting would have been 'tentative'. That's because (C2) also validates the Cautious Transitivity principle, which has good instances but, in the ROMS reading of our TSIMs, may face counterexamples:

(Cautious Transitivity) $\{I^\varphi \psi, I^{\varphi \wedge \psi} \chi\} \vDash I^\varphi \chi$[3]

Cautious Transitivity has good instances. $I^\varphi \psi$: supposing John has won the lottery, you imagine that he has a lot of money. $I^{\varphi \wedge \psi} \chi$: supposing John has won the lottery and has a lot of money, you imagine that he is to pay substantive amounts of taxes. Thus, $I^\varphi \chi$: supposing John has won the lottery, you imagine that he is to pay substantive amounts of taxes.

Cautious Transitivity for ROMS may face counterexamples, however (or at least: I have presented these ideas in various talks, where a number of people had doubts specifically on that principle even while they were on board with the rest, and in particular with Restricted Equivalence; and some came up with putative counterexamples). The issue has to do with cases where χ easily pops to mind given ψ alone, but is only dimly related to φ; for then Cautious Transitivity, acting a bit like a Cut rule in a logical calculus, washes the bridging ψ away in the conclusion. Here's a situation suggested by Claudio Calosi, that some audiences found persuasive. $I^\varphi \psi$:

---

[3] *Proof:* suppose (a) $w \Vdash I^\varphi \psi$ and (b) $w \Vdash I^{\varphi \wedge \psi} \chi$. From (a), Success, and Adjunction we get $w \Vdash I^\varphi (\varphi \wedge \psi)$, thus, by (SX), $f_\varphi(w) \subseteq |\varphi \wedge \psi|$ and $t(\varphi \wedge \psi) \leq t(\varphi)$. Also, $w \Vdash I^{\varphi \wedge \psi} \varphi$ (from Success $\vDash I^{\varphi \wedge \psi}(\varphi \wedge \psi)$ and Simplification). By (SX) again, $f_{\varphi \wedge \psi}(w) \subseteq |\varphi|$ and $t(\varphi) \leq t(\varphi \wedge \psi)$. Thus, by (C2) $f_\varphi(w) = f_{\varphi \wedge \psi}(w)$, and $t(\varphi \wedge \psi) = t(\varphi)$ (by antisymmetry of content parthood). Next, from (b) and (SX) again, $f_{\varphi \wedge \psi}(w) \subseteq |\chi|$ and $t(\chi) \leq t(\varphi \wedge \psi)$. Therefore, $f_{\varphi \wedge \psi}(w) = f_\varphi(w) \subseteq |\chi|$ and $t(\chi) \leq t(\varphi) = t(\varphi \wedge \psi)$. Thus by (SX) again, $w \Vdash I^\varphi \chi$.

supposing I am wearing a red shirt in Pamplona, I imagine that I am being chased by bulls. $I^{\varphi \wedge \psi}\chi$: supposing I am being chased by bulls on the streets of Pamplona while wearing a red shirt, I imagine that I die on the street. But it's not the case that $I^{\varphi}\chi$: supposing I am wearing a red shirt in Pamplona, I imagine that I die on its streets. Perhaps (C2) has to go for ROMS, in spite of its usefulness.

Here's an interesting take on the situation, suggested by a reviewer of this book: perhaps there is a relevant ambiguity in how we read '$I^{\varphi}\psi$': do we read it normatively ('Supposing φ, one ought rationally to imagine that ψ'), or descriptively ('Supposing φ, one as a matter of fact imagines that ψ')? Well, my initial idea was that the TSIMs in ROMS clothing capture something concerning what we would normally do in a mental simulation exercise (hence the attention to features of ROMS taken from cognitive psychology); but this would also be much of what we ought to do: what we do in mental simulation is usually the right thing to do.

Think about the claim that the topicality constraint captures the idea that mental simulation, unlike free associations of ideas, is focused on the issue one is addressing. Supposing that Brexit causes a recession, we normally don't imagine necessary truths that have nothing to do with the issues addressed in the suppositional exercise – say, that $2 + 2 = 4$ – just because they are true anyway, and so will be true in any possible scenario. Also, we shouldn't imagine that – and so the framework represents, to that extent, what we should (not) do –, for it would clearly take us off-topic. Nor does this seem to be motivated only by considerations of cognitive economy. That seems an idle thing to imagine, with respect to the goal of the suppositional exercise, *even* when one is a computationally unbounded agent with all the time in the world. Perhaps the diverging intuitions on Cautious Transitivity just show this: that's one inferential schema for which our intuitions on what we should do, and our views on what we generally do, may diverge; and the ambiguity between describing and prescribing becomes relevant.

In Berto (2017a), I suggested that if one resorts to an extended semantics for imagination-as-ROMS operators which uses, besides possible worlds, logically impossible ones (of a non-adjunctive kind: such worlds can make true two formulas without making true their conjunction, and/or vice versa), one can have (C2) and its welcome child, Restricted Equivalence, without having Cautious Transitivity because Simplification and/or Adjunction can fail in such a framework. As one can check from the relevant footnotes, the proof of Cautious Transitivity essentially uses both, whereas the one of Restricted Equivalence doesn't. That's one circumstance in which a non-classical modal logical setting including impossible worlds may have the edge over our conservative, S5-ish modal framework. There will be more, as we are to see in the following chapter.

5.5   Relaxing Topic-Inclusion?

When I have given talks on TSIMs, it has sometimes been objected that the topicality constraint for $X^\varphi\psi$, whereby the topic of ψ must be fully included in that of φ, is too draconian and should be relaxed. This might be taken as an issue for any kind of (two-place) TSIM, but it seems to be felt especially pressing for the ROMS reading: isn't imagination, even when restrained, focused, and reality-oriented in mental simulation, supposed to fruitfully expand or take us beyond the initial topic? (I owe this to Chris Badura, Aaron Cotnoir, Tim Williamson, and others.) Hence, I decided to put the (as you will see, rather tentative) discussion of this at the end of our chapter on imagination.

I think the objection points at deep issues. But it should be disentangled from a narrow-minded view of how topics are to work (I don't mean to imply that those who raised the objection were guilty of such narrow-mindedness). The narrow-minded view comes from thinking of topic-inclusion as some kind of analyticity.

As already flagged in Section 3.1, two-place TSIMs are similar enough to operators found in systems which have been called 'tautological entailment' or 'analytic containment' logics: see for instance Parry (1933); Van Fraassen (1969); Angell (1977). But even supposing we have clear cases of analytic containment (the concept of a body analytically including that of extension, or whatever may fit the bill for you), topic-inclusion is not supposed to work like *that*. It's not the case that when $t(\varphi) \leq x$, the topic of φ is included in a given topic x, that is, φ is entirely about x, then one (perhaps an ideal reasoner) should always be able to extract the former from the latter *a priori*, via conceptual analysis, whatever this amounts to. The topic of 'Jane is a lawyer' can be included in the one of Jane's profession, but one cannot extract the former *a priori* via analysis of the concept of Jane's profession. The topic of 'Maine experiences cold winters' can be included in the topic of what New England is like (Goodman 1961), but one cannot extract the former *a priori* via analysis of the concept of New England. The topic of 'Caesar crosses the Rubicon' can be included in whatever topic is suitably associated to Caesar, but one cannot extract the former *a priori* via analysis of the concept of Caesar (whatever Leibniz really thought about this).

Where are the deep issues, then? Commenting on the topic-inclusion condition for the imaginative TSIM, Chris Badura has a nice example:

> The condition on contents is too strict, however. It is particularly problematic when the subformulas in the imagination formula are atomic. Consider the following example, where Gwenny is a dog, and Helena is concerned with taking her to the lake: 'In an act of imagining that Gwenny is at her favourite lake, Helena imagines that Gwenny swims in her (Gwenny's) favourite lake'. This expresses a perfectly legitimate imaginative episode. Moreover, the episode is not especially creative, logically anarchic, or irrational. It is an imaginative episode from everyday life. Thus, if such an episode occurs, the formal counterpart of expressing it should come out as true. And it should also come out as true for the right reasons. Problematically, Berto's account does not predict this episode as true since linguistic intuition suggests that the content of 'Gwenny swims in her favourite lake' is not a part of the content of 'Gwenny is at her favourite lake'. (Badura 2021b, 2)

The situation is one in which Helena is addressing the issue (introduced by the question): What if I take Gwenny to her favourite lake? This sets the topic of the ROMS exercise. Helena supposes that Gwenny is at her favourite lake. That Gwenny swims in there sounds fully on-topic. Now if $p =$ 'Gwenny is at her favourite lake', $q =$ 'Gwenny swims in her favourite lake', our TSIM semantics *can* make '$I^p q$' true ('Supposing Gwenny is at her favourite lake, one [Helena] imagines that Gwenny swims in her favourite lake'). The issue is that it does so in a rather unenlightening way.

Take a model on a frame with a topic-assignment whereby q turns out to be fully on-topic with respect to p, $t(q) \leq t(p)$. Let your base world be w, and say that all worlds in $f_p(w)$, that is, all worlds accessible from w given suppositional input p, make q true. Then it will be true at w that $I^p q$. Of course, $I^p q$ won't be true at all worlds of all models, and so it won't be a logical truth. But that's how it should be (nor does Badura take issues with this): that, supposing Gwenny is at her favourite lake, Helena imagines that Gwenny swims in her favourite lake, shouldn't be a logical truth, and in fact it should be a contingent proposition.

What is, I guess, unsatisfactory, is that the result is plainly delivered via the fact that the semantics imposes no constraints on how topics are assigned to atomic formulas. It just so happens that the topic of 'Gwenny swims in her favourite lake' turns out to be included in that of 'Gwenny is at her favourite lake'. We – from the outside, so to speak – can make sense of it by knowing that the situation is one in which Helena addresses the issue of Gwenny's possible lake-related activities, which sets the topic for the suppositional exercise. But the semantics doesn't quite represent this.

Badura thinks that the problem lies with the fact that our $\mathcal{L}$ is a merely propositional language, where logically atomic sentences like 'Gwenny is at her favourite lake' are represented by sentential variables. If we want to represent topicality connections between atomic sentences that depend on their subsentential components, we will need to move to a first-order language, unpack the subsentential structure of sentences, and examine how topicality works there. Badura does precisely that: he relies on the theory of topics proposed in Hawke (2018), and expands it to a first-order language with modal operators as well. You may want to look at Chris' paper yourself. Here are some general remarks on why the move to first-order is bound to be complicated, aside from the specifics of his approach.

In a variety of contexts, and depending on one's favourite approach to subject matters, the topics of whole sentences may take us far away from the semantic values (senses, denotations, or whatnot) of their sub-sentential components.I suspect this is what, as I flagged a couple of times already, has made subject matter theories progress more slowly with subsentential items than with whole sentences. I think, e.g., that sect. 6 of Yablo (2018)'s reply to Kit Fine (2020)'s review of *Aboutness* points at this.

We have seen in Subsection 2.2.1 that the topic of atomic sentences does not seem to plainly supervene on the semantic values of their subsentential constituents. There, we rejected Constituent Equivalence, the idea that aRb and bRa always get the same topic, following the Yablovian and Finean thought that subject matter should concern, not only the things one talks about, but also what one says about them (the 'Dog bites man' vs 'Man bites dog' example). Merely object-oriented accounts of subject matters, whereby the subject matter of a sentence boils down to the semantic values of the subsentential constituents taken together, have been convincingly criticized, e.g., in Yablo (2014), sect. 2.1, as well as in Hawke (2018).

Does topic supervene on the subsentential constituents plus the way they are arranged? Going back to an example from Subsection 2.2.3: 'Matt is a *communist*' and '*Matt* is a communist' have the same words arranged in the same order. However, the difference in focus may be enough to mark a difference in topic: see Plebani and Spolaore (2021) on the links between focus and subject matter. If one has a question-based approach to topics, whereby the subject matter of φ in a context is linked to the question φ can be taken as answering to in that context, the two Matt sentences can be seen as differing in topic insofar as they can answer to different questions: the former can be an answer to 'What are Matt's political views?', the latter to 'Do you know any real-life communist?'.

As already remarked in Section 2.2: that, on the one hand, topics can be associated with questions and on the other, any old thing or state of affairs may seem capable of making for a conversational topic, may pull in different directions. It might be that more object- or states-of-affairs-oriented accounts of subject matter tie the topic of a sentence more closely to the semantic values of its subsentential constituents; whereas more question-oriented accounts allow more freedom. As a consequence, they are more liberal also when it's about which topic-inclusions should hold. One could even legitimately, if rather indirectly, reply to such a question as 'Does Matt believe in God?' by saying 'Well, Matt is a communist', appealing to the commonsensical wisdom that communists are godless. Some approaches to topics can be *very* context-dependent and disconnected from the subsentential constituents of sentences.

So how about tampering with topic-inclusion while holding fixed our propositional language $\mathcal{L}$ and (the rest of) the semantics as we have them now? One may propose that we relax topic-inclusion to topic-overlap: for $I^\varphi \psi$ to be true, we only ask that φ and ψ share subject matter (given topic parthood or inclusion, $\leq$, one can define overlap between topics, '∘', the standard mereological way, as: $x \circ y := \exists z (z \leq x \land z \leq y)$). In an episode of ROMS, one can progress beyond the topic of the initial supposition, provided the move involves no jump to a topic that does not even overlap with the starting supposition.

This, however, will turn our conditional-like TSIMs into something like 'ordinary' (well ...), variably strict relevant conditionals satisfying (a non-syntactic variant of) the Variable Sharing Property from relevant logic (recall Section 3.3). We get Addition: $I^\varphi \psi$ now logically entails $I^\varphi (\psi \lor \chi)$. We cannot claim anymore to model agents who really keep their thinking focused in exercises of ROMS: when, supposing Stauffenberg puts the bomb on the other side of the table, they imagine that Hitler gets killed, they now automatically imagine that either Hitler gets killed, or there's life on Kepler-442b. This sounds bad: the imaginative episode was not supposed to be about Kepler-442b at all.

In the broader-than-ROMS TSIM setting, we lose other nice features brought by the full topic inclusion constraint. The failure of topic-expanding inferences like Addition helped us to model, recall, agents with a limited conceptual repertoire, who may fail to X that φ even when φ is one elementary logical step away from what they X because the topic of φ is alien to them. Also, in the KRI setting, we will have agents such that, when their (empirical) information φ puts them in a position to know that $\psi =$ 'They have hands', it automatically puts them in a position to know that $\psi \vee \neg\chi$, and so that $\neg(\neg\psi \wedge \chi) =$ 'It's false that they are handless brains in a vat', against fallibilist insights, etc. (See the story in Section 3.3.)

A better idea to relax the topic-inclusion constraint has come from Aaron Cotnoir. Instead of requiring that the topic of ψ be included in that of φ for $X^\varphi\psi$ to be true, we require that the topic of ψ be included in the *topological closure* of that of φ. Imaginers can jump outside of the starting suppositional topic. To keep the exercise focused, it's enough that they only jump to nearby, suitably connected topics. Topological connectedness may help.

Here's a sketchy presentation of how it may work. We add to our frames for $\mathcal{L}$ a topological closure operator $f : \mathcal{T} \to \mathcal{T}$ which satisfies, for all $x, y \in \mathcal{T}$:

(i) $x \leq f(x)$ [Inclusion]

(ii) $f(x) = f(f(x))$ [Idempotence]

(iii) $f(x \oplus y) = f(x) \oplus f(y)$ [Additivity]

Technically, (i)-(iii) are the so-called Kuratowski axioms, making of f a Kuratowski closure operator on the partially ordered set $(\mathcal{T}, \leq)$. We can also use overlap $\circ$ and the f operator to define a relation C of connectedness between topics: $xCy := x \circ y$ *or* $x \circ f(y)$ *or* $f(x) \circ y$. (We can also define further topological notions, for example: topic x is *externally connected* to topic y iff xCy & $\sim x \circ y$. *Self-connectedness* for topic x goes thus: $\forall y \forall z(x = y \oplus z \Rightarrow yCz)$.)

Now here's Aaron's idea: the topic of a given suppositional input φ in an act of ROMS can be expanded to other, distinct, but connected topics in the imaginative output. The expansion, however, is regimented: only topics closely related to the input will be considered. That is, the topic of ψ must be within the closure of the topic of input φ. The refurbished truth clause for the TSIMs (in the set-selection function notation) goes thus:

$$(\mathrm{S}X')\ \ w \Vdash X^\varphi\psi \Leftrightarrow [1]\ f_\varphi(w) \subseteq |\psi|\ \&\ [2]\ t(\psi) \le f(t(\varphi))$$

The Kuratowski principles make intuitive sense of [2]. Inclusion (i) guarantees that the closure $f(x)$ of a given topic x will always be an expansion: it will enlarge the original one, but never take us far away from it. Hence, as for the TSIMs, the topic of ψ will always be connected, and sometimes only externally connected, to that of φ, but never completely unconnected. Idempotence (ii) guarantees that one cannot repeat the expansion unless the input changes. So for example, one can take an input, expand the topic in an exercise of ROMS, and then once in a new topic find new inputs and go again. But absent new inputs, the possible outputs of imaginings cannot change. Additivity (iii) guarantees that closing the topic of φ is the same as closing the topics of the atoms in φ and then fusing them: expansions of the whole never outstrip the expansions of the parts.

This setting may improve the mood of those who complain that imagination should be ampliative, and so one who supposes φ should be allowed to go beyond the topic of φ, while retaining the focus of ROMS. There are reasons, thus, for liking the addition of the Kuratowski operator. We will see that the operator can be put to good use in our topic-sensitive and probabilistic approach to indicative conditionals, in chapter 8, in particular in Section 8.3. In that setting, it will help to make acceptable a number of indicatives, in contexts where the topic of their consequent may not seem to be directly included in the topic of their antecedent.

However, the refurbishment $(\mathrm{S}X')$ makes little difference for the *logic* of the TSIM operators. Indeed, Aybüke Özgün has proved, in a joint paper with Aaron (Özgün and Cotnoir 2021), that any formula of $\mathcal{L}$ which turns out to be logically valid in the old setting is also valid in the new setting with the topologic-sensitive TSIM operator, and vice versa. In other words: adding the Kuratowski closure operator to our frames, and relaxing topic-inclusion for our TSIMs as per (clause [2] of) (SX'), does not change the logic. So, whereas the semantics using the closure operator may be philosophically more satisfactory, giving us a sense of *how* subject matters can be suitably expanded without going off-topic, e.g., in an act of focused ROMS, it does not change one bit the logic of ROMS and, more generally, the logic of two-place TSIMs investigated in these chapters. I take this to be a good result for my simple setting, in spite of doubts on the draconian nature of plain topic-inclusion.

5.6   Chapter Summary

This chapter has explored a reading of the two-place TSIM operators, in which they are interpreted as capturing a certain kind of imaginative exercise, ROMS: the activity one engages in, when one supposes that something is or had been the case and wonders what else will be or would have been the case in the hypothetical scenario. It has argued that imagination, of this sort, can have epistemic value insofar as its departure from reality is regimented and only partly voluntary, and has connected ROMS to the formation of conditional beliefs. The chapter has also explored the idea of equivalence in imagination: what it means that φ and ψ play the same cognitive role in one's mental life, in particular when one engages in suppositional thinking. It has discussed the suggestion that topic-inclusion may be too draconian a constraint for our TSIMs, at least (and perhaps not only) in the ROMS setting.

6

Hyperintensional Belief Revision

Constraints (C1)-(C4) from Section 3.5 are automatically satisfied by adding to the basic semantics of chapter 3 a function, \$, assigning to each w a finite set of nested subsets of W, which work similarly to the 'spheres' of Lewis (1973)'s classic semantics for counterfactuals: $\$(w) = \{S_0^w, S_1^w, ..., S_n^w\}$, with $n \in \mathbb{N}$, such that $S_0^w \subseteq S_1^w \subseteq ... \subseteq S_n^w = W$. Next, for each $\varphi \in \mathcal{L}$ and $w \in W$, $f_\varphi(w)$ goes thus: if $|\varphi| = \emptyset$, then $f_\varphi(w) = \emptyset$. Otherwise, $f_\varphi(w) = S_i^w \cap |\varphi|$, where $S_i^w \in \$(w)$ is the smallest sphere such that $S_i^w \cap |\varphi| \neq \emptyset$. Because the number of spheres around each w is assumed to be finite, the system satisfies Lewis' Limit Assumption: the existence of a smallest such S_i^w for each $w \in W$ and $\varphi \in \mathcal{L}$ is automatically guaranteed. For a proof that the semantics satisfies (C1)-(C4), check Priest (2008), 91-2.

Read the corresponding $wR_\varphi w_1$ as saying that w_1 is one of the most plausible worlds where φ holds, given the system of beliefs of the agent located at w. ('Plausible worlds' is shorthand for: worlds representing possibilities which look plausible in light of the agent's beliefs. Plausibility comes in degrees.) We then relabel our generic TSIM '$X^\varphi \psi$' as '$B^\varphi \psi$', for we take it as expressing conditional belief, or disposition to revise beliefs, or (static) belief revision: we read it as 'Conditional on φ, one believes that ψ', or as 'If one were to get the information that φ, one would believe that ψ was the

Topics of Thought: The Logic of Knowledge, Belief, Imagination.
Francesco Berto, Oxford University Press. © Francesco Berto 2022.
DOI: 10.1093/oso/9780192857491.003.0006

case'. We talk of conditional belief, not 'conditional credence', as the latter terminology seems more often associated to a graded or probabilistic notion: see, e.g., Leitgeb (2017), 9-10. As in the Grove (1988) doxastic-epistemic logic reformulation of the Lewisian insight, we don't demand that $w \in S_0^w$, that is, that the base world be in the innermost sphere: in Lewis' terminology, we have a system of spheres which is not even 'weakly centered'. That's because our spheres do not model objective world-similarity, as it would be in the Lewis semantics for counterfactuals, but subjective world-plausibility, or belief entrenchment. The closeness or remoteness of worlds represents, not how objectively similar or dissimilar (in the relevant respects) they are to the base world w, but how plausible the possibilities represented by the worlds look for the believing agent located at w. The innermost sphere at the core, S_0^w, gives the most plausible worlds for the agent; w itself need not be among the innermost worlds (intuitively: the agent may have false beliefs).

With this set-up, Section 6.1 introduces the AGM belief revision postulates and comments on frameworks which aim at recapturing AGM in a modal-epistemic logic setting. Section 6.2 examines which AGM principles have TSIM-counterpart validities, and which don't. It also discusses how the TSIM setting, in conditional belief clothing, relates to principles of non-monotonic logic. Section 6.3 shows that TSIM belief revision is not trivialized by the receipt of inconsistent information. Section 6.4 hints at how to develop the framework in the direction of a properly dynamic belief revision, broadly in the style of Dynamic Epistemic Logic. In particular, in a topic-sensitive dynamic setting one can have a dynamics involving the topics themselves, representing how agents can come to grasp new ones.

6.1   The Non-Hyperintensionality of AGM

The celebrated AGM approach to belief revision, due to Alchourrón, Gärdenfors, and Makinson (Alchourrón et al. 1985), came from insights originally about theory revision. The idea was to model how a scientific theory gets minimally altered to explain phenomena it did not account for before (Van Ditmarsch et al. 2008, 44).

AGM does not explicitly include belief operators in the object language. It features, instead, sets of formulas of the language closed under (classical) logical consequence, called *belief sets*, and axioms regulating the operations of expansion (+), contraction (-), and revision (*) on the sets. Expansion is about adding a formula φ to belief set K; contraction is about subtracting it, in the sense that the set does not entail φ (anymore). Revision, on which we focus, is about minimally changing the belief set to accommodate φ. Roughly: φ is added to K while other formulas are taken away so that the resulting belief set is consistent. Less roughly, revision is captured by a bunch of axioms. Using the AGM notation '$K * \varphi$' (belief set K after revision by φ), the axioms, with common names from the literature, are

(K*1) $K * \varphi$ is a belief set. [Closure]

(K*2) $\varphi \in K * \varphi$. [Success]

(K*3) $K * \varphi \subseteq K + \varphi$. [Inclusion]

(K*4) If $\neg\varphi \notin K$, then $K + \varphi \subseteq K * \varphi$. [Vacuity]

(K*5) $K * \varphi = K_\perp$ if $\vdash \neg\varphi$. [Triviality]

(K*6) If $\vdash \varphi \equiv \psi$, then $K * \varphi = K * \psi$. [Extensionality]

(K*7) $K * (\varphi \wedge \psi) \subseteq (K * \varphi) + \psi$. [Superexpansion]

(K*8) If $\neg\psi \notin K * \varphi$, then $(K * \varphi) + \psi \subseteq K * (\varphi \wedge \psi)$. [Subexpansion]

The important ones for us are (K*1), (K*2), (K*5), (K*6), because they are the ones one may be unhappy with, if one is after modelling a certain kind of believing agent. (K*1) has it that $K * \varphi$, *qua* belief set, is closed under full classical logical consequence. (K*2) says that revision succeeds, i.e., after revision by φ, this is in the belief set. (K*5) states the triviality of belief sets revised in the light of inconsistent information: if φ is a logical inconsistency (i.e., its negation is a logical theorem), then $K * \varphi = K_\perp$, the trivial belief set comprising all formulas of the language. (K*6) requires that, if φ and ψ are logically equivalent (their material equivalence is a theorem), then $K * \varphi = K * \psi$, that is, revising by either gives the same belief set.

Agents who revise beliefs as per the axioms are fully logically omniscient. While work on belief *bases* (belief sets not closed under logical consequence, see Hansson (1999)) has gone a long way towards reducing the idealization in the original AGM approach, this is still very much present in static and dynamic doxastic and epistemic logics, which aim at recapturing AGM within a modal language with appropriate formal semantics and proof theory. The reason is, plainly, that such approaches build their conditional belief or belief revision operators on top of a largely standard, normal modal framework.

Works such as Spohn (1988); Segerberg (1995); Lindström and Rabinowicz (1999); Board (2004); Van Ditmarsch (2005); Asheim and Sovik (2005); Leitgeb and Segerberg (2005); Van Benthem (2007); Van Ditmarsch et al. (2008); Baltag and Smets (2008b); Van Benthem (2011); Girard and Rott (2014), include operators for conditional belief and/or for dynamic belief revision, say $[*\varphi]B\psi$ ('After revision by φ, it is believed that ψ'), which closely mirror the original AGM postulates. As counterparts of (K*6), in particular, such logics will typically have principles like:

From $\varphi \equiv \psi$, infer $B^\varphi \chi \equiv B^\psi \chi$

From $\varphi \equiv \psi$, infer $[*\varphi]\chi \equiv [*\psi]\chi$

Such modal operators are, thus, merely intensional: incapable of detecting differences more fine-grained than ordinary modal ones.

One may want to model, instead, agents whose belief states, against (K*1), fail to be closed under full classical logical consequence. Against (K*5), one may want to model agents who do not trivially believe everything just because they occasionally hold inconsistent beliefs: after all, we don't go crazy just because we can be, as we occasionally are, exposed to inconsistent information and come to hold inconsistent beliefs on this basis. (There can be such information, even in the unfortunate case that dialetheism is mistaken, if information needn't be factive: see Section 4.2.) Against (K*6), one may want to model agents who hold different beliefs conditional on logically or necessarily equivalent pieces of information, or who are disposed to revise their beliefs differently when learning logically or necessarily equivalent things.

One can achieve some of these results in doxastic-epistemic logics which resort to Scott-Montague neighbourhood semantics (Scott 1970; Chellas 1980; Pacuit 2017). The truth conditions for the belief operator in such semantics have it that $B\varphi$ (the agent believes that φ) is true at w iff $|\varphi| \in N(w)$, the (thin) proposition that φ belongs to the neighbourhood of w: a set of sets of worlds assigned to w by the neighbourhood function N, which, intuitively, lists for each world the set of (thin) propositions necessary at it. Because the neighbourhood set can display little logical structure, such belief operators defy most closure features: agents are not modelled as believing all logical or necessary truths and all consequences of their beliefs. Such semantics have thus been used to provide (dynamic) epistemic logics for realistic agents (Balbiani et al. 2019, to be discussed in the next chapter), and also to model allegedly logically anarchic intentional states, such as imagination (Wansing 2017).

But when φ and ψ are intensionally equivalent, having the same worlds in their truth sets, they will inevitably coincide in neighbourhoods and as a result, belief in either will automatically entail belief in the other. So even neighbourhood-based approaches don't deliver the desired hyperintensionality:

1. $7 + 5 = 12$.

2. Extremally disconnectedness is not a hereditary property of topological spaces.

3. Bachelors are unmarried.

4. Baryons are hadrons with odd numbers of valence quarks.

5. Socrates exists.

6. {Socrates} exists.

These are pairwise intensionally equivalent, necessary (of the same kind of necessity: mathematical for (1)-(2) and, say, definitional for (3)-(4)) or co-necessary ((5)-(6)). However, we may want to represent agents who believe, or come to believe, only the odd items in the pairs; or who believe, or come to believe, different things given the odd and given the even items.

One way to fix this within a neighbourhood approach (suggested by Hannes Leitgeb) would be to change the set of worlds one has in the semantics: instead of having only absolutely or (broadly) metaphysically possible ones, we allow worlds that represent metaphysical, mathematical, or definitional impossibilities. Then one can split neighbourhoods in a more fine-grained way, by having, e.g., mathematically impossible worlds where 7 + 5 still equals 12, but topological spaces have bizarre features. However, this may not be enough to tell logical equivalents apart; for this, one would need full-fledged logically impossible worlds, too. Now while I have explored the impossible worlds approach at length myself (Berto and Jago 2019), the TSIM setting brings the benefit that, by representing the topicality component of the contents of attitudes separately, it allows one to smoothly focus on modelling agents with *conceptual* limitations, taking these as limitations in the subject matters one is positioned to grasp. One may, for instance, be on top of elementary arithmetic while having little competence in topology; or be fluent enough in English to grasp the meaning of 'bachelor', while having never heard of notions from particle physics; or have no trouble with talk of existence for concrete objects, while having no idea what a set is. As already argued since Section 3.3, one may want to consider agents who have limitations of this kind in their conceptual repertoire in spite of being unbounded in other respects, e.g., deductively. TSIMs are especially good with this.[1]

---

[1] We are in the vicinity of the framing effect, which involves the having, or coming to have, different attitudes, and in particular beliefs, on the basis of logically or necessarily equivalent pieces of information. When I wrote Berto (2018), on which parts of this chapter rely, I thought the setting proposed (t)here would be good enough to model certain typical forms of framing. In the cases of framing most discussed in cognitive psychology, choice theory, economics, etc., the agents are on top of all the involved propositional contents and, in

6.2   TSIMs and the AGM Postulates

By giving us constraint (C1), the total plausibility preordering of worlds guarantees that our conditional belief operator
satisfies the principle corresponding to the Success postulate
of AGM:

(Success) $\vDash B^\varphi \varphi$

Success guarantees that, conditional on φ, one does believe
φ, or that if one were to get the information that φ, one
would believe that φ was the case. What looks as perhaps
the most obvious AGM postulate is problematic for *dynamic*
belief revision (of which we will talk in the final section of this
chapter), due to such phenomena as the Moore sentences ('φ,
but I do not know/believe that φ'). But it's generally taken as
unproblematic for conditional belief or static belief revision:
Van Benthem and Smets (2015) provide a nice discussion.

Simplification, Adjunction, and Commutativity keep holding, as per the basic semantics from chapter 3 – and as
they are supposed to do for a conception of belief, not as
mere syntactic mental symbol tokening but as a properly
intentional mental state endowed with contents, and directed
towards the situations those contents are about: remember
Section 3.2. I argued there that, in particular, Adjunction
is justified for all-or-nothing belief, and basically referred
you to Hannes Leitgeb (2017)'s stability theory of belief
on how one could make this cohere with degrees of belief
and the passing of intermediate thresholds in probabilistic or
degree-theoretic settings. Leitgeb's work, as mentioned there,
addresses Lottery-Paradox-like situations by introducing vari-
ations on the threshold across epistemic contexts.

Now Leitgeb takes the conservative stance of understanding
the contents of attitudes as plain thin propositions: sets of

---

particular, subject matters. For instance, they may believe that they should
apply for the job when told they have 40% chances of getting it, not believe
that much when told they have 60% chances of failing, although they are on
top of the relevant probabilistic notions and they are also, in a sense, aware
that 40% success and 60% failure are necessarily equivalent. I now think such
typical framing may be modelled in a more accurate way by the TSIM setting
of the following chapter 7, where that 'in a sense' is made more precise.

possible worlds. But he also hints at how all-or-nothing belief, with the context-sensitivity induced by the stability view, could be understood hyperintensionally. Epistemic contexts are modelled as partitions of modal space induced by a salient or relevant issue or question the agent is tackling. As we know since our chapter 2, this is also one common way of understanding topics. Once one embeds it into the propositional contents themselves, thereby making belief hyperintensional, Leitgeb's context-sensitivity starts to look similar enough to our topic-sensitivity of belief:

> In spite of the attractions of the standard (possible worlds or neighbourhood) semantics of belief, perhaps belief contents ought not really to be identified with sets of possible worlds after all but instead with more fine-grained entities, so that one might rationally believe that A, not believe that B, while 'A' and 'B' are true precisely in the same worlds. [...] In that case, changing a partition *would* affect a belief content: X under partition π would differ in content from X under partition π', if the partition co-determines content. [...] What one believes also depends on what the belief is about. (Leitgeb 2017, 142)

By giving us constraint (C2), the plausibility preordering also ensures the validity of Cautious Transitivity (the proof is as per the relevant footnote in Section 5.4) and Cautious Monotonicity:

(Cautious Transitivity) $\{B^\varphi \psi, B^{\varphi \wedge \psi} \chi\} \vDash B^\varphi \chi$

(Cautious Monotonicity) $\{B^\varphi \psi, B^\varphi \chi\} \vDash B^{\varphi \wedge \psi} \chi$[2]

These two plus Success make our conditional belief TSIM comply with Gabbay (1985)'s minimal non-monotonic validities, mentioned in Section 4.6, and which failed for KRI.

---

[2] *Proof:* suppose (a) $w \Vdash B^\varphi \psi$ and (b) $w \Vdash B^\varphi \chi$. From (a), Success ($\vDash B^\varphi \varphi$), and Adjunction, we get $w \Vdash B^\varphi(\varphi \wedge \psi)$, thus by (SX), $f_\varphi(w) \subseteq |\varphi \wedge \psi|$. Also, $w \Vdash B^{\varphi \wedge \psi} \varphi$ (from Success $\vDash B^{\varphi \wedge \psi}(\varphi \wedge \psi)$ and Simplification), so by (SX) again, $f_{\varphi \wedge \psi}(w) \subseteq |\varphi|$. Then, by (C2), $f_\varphi(w) = f_{\varphi \wedge \psi}(w)$. From (b) and (SX) again, we get $f_\varphi(w) \subseteq |\chi|$, thus $f_{\varphi \wedge \psi}(w) \subseteq |\chi|$. Also, $t(\chi) \leq t(\varphi) \oplus t(\psi) = t(\varphi \wedge \psi)$. Thus, $w \Vdash B^{\varphi \wedge \psi} \chi$.

However, $B^\varphi\psi$ fails (the counterparts of) two more, which hold in Kraus et al. (1990)'s mainstream system C of non-monotonic implication (say '$|\sim$'); These are often called Left Logical Equivalence (If φ and ψ are logically equivalent, and $\varphi \mathrel{|\sim} \chi$, then $\psi \mathrel{|\sim} \chi$) and Right Weakening (If $\varphi \mathrel{|\sim} \psi$, and ψ logically entails χ, then $\varphi \mathrel{|\sim} \chi$).

The counterpart of Left Logical Equivalence is just Equivalence from Section 3.4, which fails as per the basic semantics. Against AGM's (K*6), one can believe different things given (topic-divergent) logically or intensionally equivalent inputs:

$$\{B^\varphi\psi, \varphi \succ\!\!\prec \chi\} \not\models B^\chi\psi$$

One believes that one could have a chat with Socrates, conditional on Socrates' existing; one doesn't believe that much, conditional on the existence of the singleton of Socrates.

But, we can of course have our *restricted* counterpart of (K*6), thanks to constraint (C2). This is our old Restricted Equivalence principle from Section 5.4, now relabeled for belief:

(Restricted Equivalence) $\{B^\varphi\psi, B^\psi\varphi, B^\varphi\chi\} \models B^\psi\chi$

Remember that Restricted Equivalence was marketed back then as a principle of cognitive equivalence. This still holds here: when φ and ψ play the same role in our cognitive life, they *will* be interchangeable in our conditional belief and belief revision policies: one believes that John has no marriage allowance, both conditional on John being a bachelor and conditional on John being unmarried.

The counterpart of Right Weakening is the usual TSIM failure of Closure under Strict Implication, which, again, happens due to $\prec$ failing to be topic-preserving:

$$\{B^\varphi\psi, \psi \prec \chi\} \not\models B^\varphi\chi$$

Conditional on information on Socrates' recent, lively speech, one believes that Socrates (still) exists. There is no way for Socrates to exist without {Socrates} existing, but one has just no idea what sets are.

Things are different if, conditional on the same things, one *believes* the strict implication itself (compare the COOKIT principle for KRI from Section 4.5). This principle, which might be called Closure under Believed Implication, is valid:

(CBI) $\{B^\varphi\psi, B^\varphi(\psi \prec \chi)\} \vDash B^\varphi\chi^3$

Here both ψ and $\psi \prec \chi$ are believed conditional on the same φ. Thus, the agent is on top of all the relevant subject matters. Then the agent also believes that χ, conditional on the same φ. The final proviso is essential: given the non-monotonicity of B, the entailment does not hold anymore if the conditional input is allowed to change across the involved formulas.

6.3   Revising by Inconsistent Information

TSIM belief revision is not trivialized by incoming inconsistent information. We already know from chapter 3 that, thanks to their topic-sensitivity, our TSIMs have an element of relevance in the sense of relevant logics. Our S5 strict implication is explosive, $\vDash (\varphi \wedge \neg\varphi) \prec \psi$ (trivially: for all w, $w \nVdash \varphi \wedge \neg\varphi$). But, against AGM's (K*5), the following ensures that we do not believe arbitrary, irrelevant things conditional on explicitly inconsistent information:

$$\nvDash B^{\varphi \wedge \neg\varphi}\psi^4$$

Although there is no possible world where a contradiction is true (unless dialetheism is right, that is), inconsistent information may still be about something, without being about everything. In general $\varphi \wedge \neg\varphi$ is about what φ is about, and this may not include the content of ψ ('Snow is white and not white' is about snow's being white, or the colour of snow, etc. – not about grass' being purple). In our topic-sensitive setting, one may say, there is a difference between

---

[3] *Proof:* let $w \Vdash B^\varphi\psi$ and $w \Vdash B^\varphi(\psi \prec \chi)$. By the former and (SX), for all w_1 such that $wR_\varphi w_1$, $w_1 \Vdash \psi$, and $t(\psi) \leq t(\varphi)$. By the latter and (SX) again, for all w_1 such that $wR_\varphi w_1$, $w_1 \Vdash \psi \prec \chi$, thus given (S$\prec$) for all $w \in W$, if $w \Vdash \psi$ then $w \Vdash \chi$. So in particular for all w_1 such that $wR_\varphi w_1$, $w_1 \Vdash \chi$. Also, $t(\psi \prec \chi) = t(\psi) \oplus t(\chi) \leq t(\varphi)$, thus $t(\chi) \leq t(\varphi)$. Thus by (SX), $w \Vdash B^\varphi\chi$.

[4] *Countermodel:* let $W = \{w\}$, $t(q) \nleq t(p)$. $|p \wedge \neg p| = \emptyset$, thus $f_{p \wedge \neg p}(w) = \emptyset \leq |q|$. However, $t(q) \nleq t(p \wedge \neg p) = t(p) \oplus t(\neg p) = t(p)$. Thus, by (SX), $w \nVdash B^{p \wedge \neg p}q$.

being informationally trivial, and being topically trivial, or being about whatever. I'll come back to this distinction in Section 6.4 below.

For similar reasons, conditional on φ one does not automatically believe logical validities which are irrelevant with respect to φ, e.g.:

$$\nvDash B^\varphi(\psi \prec \psi)$$

$$\nvDash B^\varphi(\psi \vee \neg\psi)^5$$

Here's where non-classical logics get a revenge, though, and where a very general limitation of the TSIM framework pops up: even if our TSIMs are not explosive, they do satisfy 'small explosion' principles (I think I owe the terminology to Jorge Ferreira), like

$$\vDash B^{\varphi \wedge \neg\varphi \wedge \psi}\neg\psi$$

For $\varphi \wedge \neg\varphi \wedge \psi$ is true nowhere, and topicality is preserved here. In the conditional belief reading of TSIMs, we believe conditional on φ all classical entailments of φ which are also topic-preserving. For instance, conditional on snow's being white and not white and grass being green, one also believes that grass is not green. This consequence of immanent closure may sound bad: it seems that we still believe too much! (The Parry (1968, 1989) system of analytic entailment has a similar validity: Harry Deutsch (1979) criticizes it as a 'fallacy of making too much out of a small, if nasty, mistake' (139).)

There seems to be little room for manoeuvre here, insofar as we stick to the classical and normal modal S5-ish setting I put in place since chapter 3 for our TSIMs. A framework expanded to include non-normal or impossible worlds where a contradiction can be true, thus worlds which are logically impossible (unless dialetheism is right) would help against such small detonations. I have used such a framework to model intentional operators, e.g., in Berto (2014, 2017b). Impossible worlds can also mimic topicality or relevance

---

^{5} *Countermodel:* (I do the former, the latter is analogue): let $W = \{w\}, t(q) \nleq t(p)$. Then although (trivially) $f_p(w) \subseteq |q \prec q|$, $t(q \prec q) = t(q) \oplus t(q) = t(q) \nleq t(p)$. Thus by (SX), $w \nVdash B^p(q \prec q)$.

constraints without having in the semantics an algebra of topics. What to make of this? Should we conclude that, insofar as impossible worlds semantics can deliver the same (putatively) desirable validities and invalidities as the TSIM setting, but also address some of the shortcomings of the approach developed in this book, the impossible worlds setting is better overall? I have no idea. In Berto (2021), I tentatively proposed that impossible worlds semantics wins in flexibility but loses in naturalness: validities often come only by manually adding constraints to the semantics, which look a bit artificial, or ad hoc; but I came up with nothing conclusive. Instead, I will conclude the chapter by hinting at how to have a dynamic, topic-sensitive hyperintensional belief revision setting, in the following section.

6.4   Going Dynamic: Grasping New Topics

Dynamic logics have been used extensively in artificial intelligence, e.g., for program verification (Troquard and Balbiani 2019). They include operators that represent the performing of actions (e.g., carried out by an AI agent or a software). if a is an action, one can have in the formal language a corresponding operator, say '$[a]$', and '$[a]\varphi$' says that after action a has been carried out, φ is the case. The semantics for such operators is given in terms of model-transformations: '$[a]\varphi$' is true in the initial model $\mathfrak{M}$ iff φ is true in the model, or models, obtained by transforming $\mathfrak{M}$ according to the instructions encoded in action a.

Dynamic Epistemic Logic (DEL) (Segerberg 1995; Baltag and Solecki 1998; Van Ditmarsch et al. 2008; Van Benthem 2011, etc.), focuses on epistemic actions, such as incorporating the new and truthful information that φ, or augmenting the plausibility of φ, or bumping up its probability (Baltag and Smets 2008a), also in a multi-agent setting. Such actions are represented in a more efficient and powerful way than what can be done in epistemic logic in a standard 'static' setting such as the basic Hintikkan one. As Baltag and Renne have it: 'The advantage of the dynamic perspective is that we can analyze the epistemic and doxastic consequences of actions such as public and private announcements without having to "hard wire" the results into the model from the start.' (Baltag and Renne 2016, introduction).

We can have a TSIM representing dynamic and topic-sensitive belief revision by extending, for a start, our language $\mathcal{L}$ with a dynamic operator, say, $[*...]$: if φ and ψ are well-formed formulas, so is $[*\varphi]\psi$. One can read it as 'After revision by φ, ψ is the case', but a less terse reading highlights topic-sensitivity: 'After the agent has received information φ and has come to grasp the topic of φ, ψ holds'.

Getting the semantics for this right requires an innovative, two-component model-transforming apparatus. Luckily, Aybüke Özgün has come up with one in a paper we have written together (Özgün and Berto 2020), which was in turn inspired by previous work with Peter Hawke (Hawke et al. 2020). Aybüke has paired standard DEL model update techniques with a dynamics of topics. She has added to a setting essentially like the one of the TSIM models of Section 3.1 (plus total plausibility preordering of worlds) a designated $\mathfrak{b} \in \mathcal{T}$: the topic the overall belief state of the agent is about, which represents the totality of subject matter the agent has already grasped at a given stage.

Next, for $[*\varphi]\psi$ to hold at w in a model $\mathfrak{M}$ of this kind, ψ must hold at w in the model $\mathfrak{M}^{*\varphi}$, transformed across two components. These are (1) the worldly component: all the φ-worlds have become more plausible than all the $\neg\varphi$-worlds, while the order has remained the same within the two sets; this is the normal so-called 'lexicographic upgrade' operation of DEL, whose workings are described, e.g., in Van Benthem and Smets (2015), 334-5; and (2) the topicality component: the new designated topic $\mathfrak{b}^{*\varphi}$ is now $\mathfrak{b} \oplus t(\varphi)$, representing how the agent has grasped the topic of φ, adding it to $\mathfrak{b}$.

The dynamic operator is hyperintensional: in particular, even if φ and ψ are intensionally equivalent, $[*\varphi]\chi$ may hold while $[*\psi]\chi$ doesn't when χ features a belief ascription. One can therefore come to believe different things after dynamically revising with (topic-diverging) intensionally equivalent pieces of information φ and ψ: $[*\varphi]B\chi$ can hold while $[*\psi]B\chi$ doesn't.

In this setting, one can also define *plain* belief in terms of conditional belief. The standard way of doing this is by characterizing the former as belief conditional on a tautology: $B\varphi := B^\top \varphi$ (Van Benthem and Smets 2015, 325). However, we have already seen that, in a topic-sensitive environment, one can distinguish between being uninformative the way logical truths are, because they are true at all possible worlds (of all models), and being topic-trivial. In Özgün and Berto (2020), this is represented by adding to $\mathcal{L}$ a constant, $\top$, whose truth conditions are that it's (1) true everywhere throughout modal space, and (2) such that $t(\top) = \mathfrak{b}$: the constant has as a fixed topic the whole subject matter the agent is on top of. It thus differs from a logical truth: $\varphi \vee \neg\varphi$ is about whatever φ is about, which, unlike what $\top$ is about, may not be the totality of subject matter the agent can possibly think about. Then $B^\top \varphi$ differs from belief conditional on a tautology, and gives a good characterization of plain belief.

6.5   Chapter Summary

This chapter has explored a reading of the two-place TSIM operators, in which they are interpreted as expressing conditional belief, or (static) belief revision. The reading is enforced by adding to the basic TSIM semantics a total preordering of worlds, understood as representing relative plausibility. The chapter has argued that, thanks in particular to the hyperintensionality of TSIMs, this setting reduces the logical idealization of cognitive agents in the standard AGM belief revision theory as well as in non-hyperintensional modal logics of belief revision. The chapter has concluded by hinting at a way to expand the setting into a dynamic framework, as in DEL, via a dynamics representing how the agent can come to grasp new subject matters.

7

Framing Effects

Co-authored with Aybüke Özgün

Framing effects concern one's having different attitudes towards logically or necessarily equivalent propositions (Kahneman and Tversky 1984). Framing is, thus, connected to the hyperintensionality of thought, which we know our TSIMs to be good at modelling. However, the sort of framing effects typically investigated in cognitive science, behavioural economics, decision theory, and the social sciences at large, may benefit from a bit more specificity than the kinds of hyperintensionality the TSIMs have been put to work to model so far, in particular in the hyperintensional account of belief presented in the previous chapter.

Typical framed believers are clearly logically non-omniscient. But what *kind* of non-omniscience do they display? Section 7.1 delves into this. Specifically: such believers can have different attitudes towards intensionally equivalent φ and ψ, even if they are perfectly on top of the relevant subject matters, and, in a sense, aware of the equivalence. In order to represent this, we may need more than the plain topic-sensitivity of belief in focus in the previous chapter.

Section 7.2 introduces what we take to be the required additional ingredient. A key distinction we need in order to model typical framing, we submit, is the structural one, borrowed from cognitive psychology, between beliefs *activated* in working memory and beliefs left *inactive* in long-term memory. Few proposals in epistemic logic have featured

Topics of Thought: The Logic of Knowledge, Belief, Imagination.
Francesco Berto, Oxford University Press. © Francesco Berto 2022.
DOI: 10.1093/oso/9780192857491.003.0007

formalizations of such a distinction, whereas there is an amount of literature on the distinction between explicit and implicit belief. We discuss some of it in Section 7.3. We get to our own proposal, spelling out a formal semantics, in Section 7.4. Finally, in Section 7.5 we explore its logic. In Berto and Özgün (2021) a sound and complete axiomatization is presented; we only discuss here some notable validities and invalidities.

7.1 Framed Believers

Physicians tend to believe some lung cancer patients should get surgery with a 90% one-month survival rate. Physicians tend not to believe such patients should get surgery with a 10% first-month mortality (Kahneman 2011, 367). People will believe more in a certain economic policy when its employment rate is given than when the corresponding unemployment rate is given (Druckman 2001b). Early student registration is boosted by threatening a lateness penalty more than by promising an early bird discount (Gächter et al. 2009). A good deal of behavioural economics takes its cue from framing effects. Unlike Econs, the fully consistent agents of classical economic theory who well-order their preferences and maximize expected utility, Humans can be framed: nudged into believing different things depending on how equivalent options are presented to them (Thaler and Sunstein 2008). Framing has momentous social consequences (Plous 1993; Druckman 2001a; Levin et al. 2002; Busby et al. 2018). We need a logic of framing.

We have seen since chapter 1 that our non-omniscience is tied to different, often orthogonal, features of our cognitive apparatus. This is especially relevant here. What kind of non-omniscience is involved in framing? It cannot be tied to the *a priori*/*a posteriori* distinction (as in, one believes that John is John, not that John is Jack the Ripper): that the survival rate is 90% is neither more nor less *a priori* than that mortality is 10%. Nor can it be due to computational difficulties with parsing long and syntactically complex sentences ($\varphi \supset \varphi$ vs complicated tautology): either of 'The survival rate is 90%'

and 'The mortality (rate) is 10%' is just as easy to parse as the other.

It may be that the issue is not with the nature of the attitude itself either, independently of cognitive and computational limitations. We've been exploring at length the idea that knowledge or knowability ascriptions may fail full closure in chapter 4: logically astute reasoners might fail to (be positioned to) know they are no recently envatted brains although they know they have hands, etc. We saw that the jury is out on this. What we are after now, however, is belief. When the case for knowledge not being closed under entailment even for deductively unbounded reasoners is presented, their being logically astute is usually *defined* in terms of belief: they do believe all the competently deduced logical consequences of what they know (and therefore believe), based on what they know (Dretske 1970; Nozick 1981; Holliday 2015, etc.). The open issue is whether that's sufficient for the closure of their knowledge states.

Could the kind of non-omniscience displayed by agents with framed beliefs be due to a lack of concepts, as when one believes φ but not an entailed ψ because one doesn't have some notion required to grasp ψ, and perhaps specifically the topic of ψ, what it's about? We've seen that the TSIMs are especially good with this, which gets us closer to the phenomenon we're after, but perhaps not close enough. Surely human thinkers have a limited repertoire of concepts and, as a consequence, are just unable to grasp some things language and thought in the abstract can be about; but that's not what is involved in ordinary framing. Framed physicians have all the concepts needed to fully grasp both the proposition that the survival rate is 90% and the one that the mortality is 10%. In particular, they are fully on top both of the concept *survival* and of the concept *mortality* by any conceptual or semantic competence test. They may even be aware, in a sense, that the two propositions are necessarily equivalent. Still, in some other sense, they must fail to be aware, when only the former proposition gets them to believe the patients should take surgery.

What is going on, framing theorists say (Kahneman and Tversky 1984; Kahneman 2011), is that 'The mortality is

10%', but not 'The survival rate is 90%', *makes* people think about mortality. The thought that the survival rate is 90% is not about that: on the face of it, it's about survival. Survival and death are deeply connected in anyone's mind. But, cognitively limited as we are, we may not think about mortality – and much of what comes with it – when we think about survival rates, even if we have the concept *mortality* firmly in our repertoire. We leave it asleep. In order to think that the mortality is 10%, instead, we have to think about mortality, for that's what the proposition is about.

Typically framed thinkers can have different attitudes towards necessarily equivalent propositions they perfectly grasp, due to differences in what those propositions are about, even when they are perfectly on top of the relevant topics, and even when they are, in some sense, aware of the equivalence. This is not the only way agents can be framed: *qua* psychological phenomenon, framing can involve all sorts of subtle pragmatic cues and mental associations triggered by word order, emphasis, etc. But it is a typical kind of framing, we conjecture, because it has deep roots, on the one hand, in the structure of our belief system, and on the other, in the nature of its contents. An accurate logic of framing will have to represent both roots.

7.2 Working Memory, Long-Term Memory, Aboutness

To model the structural features of our belief system responsible for typical framing, we think, one should look at a key acquisition of cognitive psychology: the distinction between working and long-term memory (Eysenck and Keane 2015, part II). (To be sure – and in reply to some helpful comments of one reviewer of this book – we don't claim to have empirical evidence of deep connections between framing and that distinction. We advance this as a conjecture, which might perhaps be operationalized and tested empirically, though we 'armchair' logical modellers have no idea of how to do it.) Researchers disagree on the nature of both kinds of memory. *Qua* logical modellers, we don't want our account to be held hostage to the next empirical discovery or consensus switch in psychological research. Luckily, we can be neutral on the more controversial issues, and just take on board the less controversial ones.

For instance, working memory (WM), which deals with the processing and short-term storage of information, is at times understood as encompassing a buffer of data at hand for the performance of cognitive tasks, plus a central executive unit: the locus of attention and cognitive control (Baddeley 1986, 2002); at times, as a plurality of modules or structures (Barsalou 1992). For our purposes, we only need to consider its most agreed-upon feature: it has limited capacity. Only a few chunks of information can be retained in WM, and only for a limited amount of time: see the views compared in Miyake and Shah (1999).

Instead, long-term memory (LTM), or the declarative part of it (Squire 1987; Schachter and Tulving 1994), is that vast knowledge base where cognitive agents store, or encode, their beliefs and knowledge about specific events (the so-called episodic memory) as well as general laws and principles (the so-called semantic memory). There's a divide in cognitive psychology, on whether WM and LTM are separate (contents are stored in LTM and retrieved from it for use in WM), or the former is just the activated part of the latter (Anderson 1983; Crowder 1993; Miyake and Shah 1999). We can be neutral on this as well.

Now, typically framed agents, we propose, can have the belief that patients should get surgery with a 90% one-month survival rate activated in their working memory, without having the intensionally equivalent belief that patients should get surgery with a 10% first-month mortality there. However, framed agents can have all the relevant information and, in particular, the concept *mortality*, in their (declarative) LTM. Let's call beliefs activated in WM *active*, and beliefs left asleep in LTM *passive*. A belief is active when it is available in WM to perform cognitive tasks with it. It is passive when it is stored, or encoded, in the agent's LTM, and left inactive there. We propose that both kinds of belief be taken as topic-sensitive. We represent them as modals in Section 7.4 below, and so they count as TSIMs.

Here are some desiderata our logic of framing should comply with. First, we evaluate ascriptions of active belief with respect to the agents' WM, and ascriptions of passive belief with respect to their LTM. Next, we may want to model realistic agents with bounded resources with respect to *both* WM *and* LTM. Psychologists contrast the limited capacity of the former with the breadth of the latter. However, neither should host all the logical consequences of what it hosts, or display an omni-inclusive conceptual repertoire. In particular, both passive and active belief must be hyperintensional: framed agents are not logically closed with respect to either.

Next, whether WM is separate from LTM, or just the activated part of the latter, no information or concept can be in WM unless it is in LTM to begin with. In particular, agents cannot have any attitude on subject matters whose concepts they simply lack. To go back to the Stalnakerian example of Section 3.3: they are as blind to them as William III was to the topic of nuclear weapons.

To get an idea of how such desiderata cooperate, consider the following two triplets of group-wise intensionally equivalent sentences:

1. $7 + 5 = 12$.

2. No three positive integers x, y and z satisfy $x^n + y^n = z^n$ for integer value of $n > 2$.

3. Extremally disconnectedness is not a hereditary property of topological spaces.

4. Triangles have three sides.

5. Bachelors are unmarried.

6. Baryons are hadrons with odd numbers of valence quarks.

(1)-(3) are necessary, of the same kind of necessity (mathematical necessity). Ditto for (4)-(6) (say, definitional necessity). Typical framed believers could find themselves in the following situation with respect to each triplet: they passively believe the first item, (1), or (4); they have the relevant information and they are on top of the basic arithmetical or geometrical subject matter involved, so it's all stored or encoded in LTM. They are just not thinking about arithmetic, or about triangles, at the moment. They actively believe the second item, (2) or (5): they have the relevant propositional content in their WM because they are currently engaged in thoughts about diophantine equations, or John's marital status. They neither actively nor passively believe the third item, (3) or (6): they just have no idea what topological spaces are and what features they have; they have never heard about exotic notions from particle physics. This three-fold distinction isn't naturally modelled in the setting of the previous chapter (compare the examples in Section 6.1).

Before we get to our own proposal to model agents of this kind, in the next section we briefly discuss some hyperintensional epistemic logics for non-logically omniscient agents already on the market, to see to what extent they *could* be used to represent framing.

7.3   Explicit and Implicit

As far as we know, few epistemic logics have aimed at directly representing the difference between WM and LTM. One distinction which may look *prima facie* similar is the one between *explicit* and *implicit* knowledge and belief, found in awareness logics developed with an eye on the logical omniscience problem (Fagin and Halpern 1988; Van Benthem and Velázquez-Quesada 2010; Velázquez-Quesada 2014): we briefly introduced and discussed them in Section 3.3. Because being unaware of φ is usually understood as not having φ present in the mind, or not thinking about φ (Schipper 2015, 79-80), the awareness approach seems especially suitable to model framing.

Remember how awareness is typically represented syntactically: one is aware of φ when φ belongs to a set of formulas, $\mathcal{A}$, the agent's awareness set. Implicit knowledge or belief are dealt with via normal Hintikka-style modal operators, whereas the corresponding explicit attitudes are defined as the combination of the implicit ones with awareness: one explicitly knows or believes that φ when one knows or believes it implicitly and φ is in the awareness set.

We mentioned in Section 3.3 that the view has been claimed to mix syntax and semantics, essentially imposing a syntactic filter over a standard Hintikkan semantics (Konolige 1986). Resorting to syntax, however, allows very fine-grained distinctions: if any bunch of sentences can serve as the awareness set $\mathcal{A}$, explicit attitudes obey no non-trivial logical closure properties. Syntactic approaches representing bodies of knowledge/belief/awareness as plain sets of sentences have then been criticized for being *too* fine-grained (Levesque 1984, 199-201). For the purposes of modelling the typical framing effects we're after, they are an overkill.

Here's why: a framed agent who actively believes $\varphi \wedge \psi$ should actively believe $\psi \wedge \varphi$, and should actively believe ψ, if our topic-sensitive view of propositional content is right. That John is tall and handsome and that John is handsome and tall are intensionally equivalent propositions, and the agent who actively believes either is already thinking about the other's topic – because it is the same topic, say, John's height and looks. That John is tall and handsome entails that John is tall, and one who actively believes the former is already thinking about the topic of the latter, as it is part of that of the former. Such mereological relations between the contents of thoughts, which have been at centre stage for much of this book, are lost in a plain awareness setting. (This doesn't rule out, we think, that plain syntactic awareness approaches may be useful in modelling some specific kinds of framing, e.g., the presentation order effects discussed in Section 3.2. As conjectured there, these may to be tied to issues with parsing the syntax of sentences.)

Nor do implicit attitudes neatly map to passive belief as implemented in LTM. Because logics featuring the explicit/implicit distinction usually take the implicit attitude as a normal Hintikkan modality, the attitude displays full logical omniscience: the agent implicitly knows or believes all logical truths, and all logical consequences of what it knows or believes. The agent has no awareness or conceptual limitations there: it is simply on top of all the relevant propositions. But, as we have remarked, LTM is not like that.

If we want to model agents who don't possess all concepts, and don't have all the logical consequences of their passive beliefs stored or encoded in LTM, passive belief should be hyperintensional, too.

Balbiani et al. (2019) present one of the few logical works with the stated aim of modelling the WM/LTM distinction. It's a powerful framework in the tradition of Dynamic Epistemic Logic, modelling the processes through which a non-omniscient agent forms its beliefs via operations of perception and inference in WM, and can store and retrieve them from LTM. Their language has an operator for explicit belief, tied to WM, and one expressing background knowledge, tied to LTM. The latter is a normal modality, and so faces the same issue as implicit knowledge in the awareness setting: the agent is logically omniscient with respect to its background knowledge.

What's more worrying for the prospects of applying the logic to framingis that explicit belief gets a Scott-Montague neighbourhood semantics (Scott 1970; Pacuit 2017): one explicitly believes that φ when φ's truth set is in the relevant neighbourhood set. We talked of the neighbourhood approach in Section 6.1, where we mentioned that it gives weak non-normal modal logics capable of breaking a number of logical closure features for their operators. In particular, one can explicitly believe a conjunction without explicitly believing the conjuncts, which, we argued above, is not good. This overkill can be fixed by adding conditions – specifically, one could close the neighbourhoods under supersets for $\wedge$-elimination: see Pacuit (2017), 81.

However, there's still the more problematic underkill we flagged in Section 6.1: even in the basic neighbourhood setting, when φ and ψ are assigned the same set of worlds as their (thin) proposition, they will be in the same sets of neighbourhoods. Thus, explicit belief in either will automatically entail explicit belief in the other. This is exactly what shouldn't happen if we want to capture framing for explicit beliefs. As we also mentioned in that section, one can play with the addition of (mathematically, logically, etc.) impossible worlds to make neighbourhood semantics more fine-grained. But the topic-sensitive approach may be better positioned to capture how the subjects of typical framing fail to (actively) think about one of two topics driving a wedge between intensional equivalents. Thus, we now move on to our own proposal and start making things formally precise.

7.4   Topic-Sensitive Active and Passive Belief

Our language $\mathcal{L}$ for this chapter will have, besides the countable set $\mathcal{L}_{AT}$ of atomic formulas p, q, r ($p_1, p_2...$), negation $\neg$, conjunction $\wedge$, disjunction $\vee$, the box of necessity $\Box$, and two belief operators, B_A and B_P. The well-formed formulas are the items in $\mathcal{L}_{AT}$ and, if φ and ψ are formulas, so are the following:

$$\neg\varphi \mid (\varphi \wedge \psi) \mid (\varphi \vee \psi) \mid \Box\varphi \mid B_A\varphi \mid B_P\varphi$$

As usual, we often omit outermost brackets and we identify $\mathcal{L}$ with the set of its well-formed formulas. Read the box as a normal epistemic or *a priori* modality (flag this: we may then see the worlds of the coming semantics as epistemically possible ones, rather than absolutely or broadly metaphysically possible ones; given that the modal is a normal one, the former will not differ that much from the latter anyway – they may, e.g., falsify narrowly metaphysically necessary claims like 'Hesperus is Phosphorus' or 'Water is H2O'); read '$B_A\varphi$' as 'One actively believes that φ', '$B_P\varphi$' as 'One passively believes that φ'. When we say something that applies to both active and passive belief, we use 'B_*'. It will come in handy to have a $\top := p \vee \neg p$ (this abbreviates a specific tautology; it is not to be confused, thus, with the $\top$ of Section 6.4) and a $\bot := \neg\top$. Again, '$\mathfrak{At}\varphi$' stands for the set of atomic formulas occurring in φ.

A *frame* for $\mathcal{L}$ is a tuple $\mathfrak{F} = \langle W, \mathcal{O}, \mathcal{T}, \oplus, t \rangle$, where W is our non-empty set of possible worlds, $\mathcal{T}$ is our non-empty set of topics, $\oplus$ is topic fusion (with topic parthood, $\leq$, defined from it as usual). The new bit is $\mathcal{O}$, a non-empty, finite subset of $P(W)$ such that $\mathcal{O} \neq \{\emptyset\}$: each non-empty $O \in \mathcal{O}$ represents the informational content of a *memory cell* (we'll come to what this is in a second).

The topic function now is $t : \mathcal{L}_{AT} \cup \mathcal{O} \to \mathcal{T} \cup P(\mathcal{T})$. It assigns a topic to each atomic formula and a non-empty, finite set of topics to each item in $\mathcal{O}$: $t(p) \in \mathcal{T}$ for all $p \in \mathcal{L}_{AT}$, and $t(O) \in P(\mathcal{T})$ is non-empty and finite for all $O \in \mathcal{O}$. Then, topics are assigned to the whole of $\mathcal{L}$ the usual way, namely with $t(\varphi) = \oplus \mathfrak{At}\varphi$, to ensure topic-transparency.

Here's what the model represents: the agent's belief system is composed of memory cells. These are chunks of LTM which can be put into (or, if one prefers, activated as) WM, that is, made available for actions of cognitive processing. A memory cell is represented by an indexed set, O_x, where $\emptyset \neq O \in \mathcal{O}$ and $x \in t(O)$. O_x is made of informational content O and topic x. Memory cells are, thus, topic-sensitive: when one is in (or activated as) WM, the agent is actively thinking about its subject matter, and has its informational content available for processing. $t(O)$ and $\mathcal{O}$ are assumed to be finite, to represent cognitive agents that can only have finitely many memory cells.

Every $O \in \mathcal{O}$ is assigned a set of topics, rather than a single topic, in order to capture the idea that the same informational content can be associated with different topics. Take our triplet of intensionally equivalent, topic-diverging sentences (1), (2), and (3) in Section 7.2. Intensional equivalence means that they have the same bunch of worlds as their truth set. Call it S. Let the topics be x, y, and z, respectively. Each of S_x, S_y, and S_z can make for a distinct memory cell, differing from the others in topic but not in informational content.

The agent's LTM is defined as:

$$LTM := \left(\bigcap \mathcal{O} \right)_{\oplus (\bigcup_{O \in \mathcal{O}} t(O))}$$

The information stored or encoded in LTM is the information available in all memory cells, taken together. The topic of LTM is the fusion of those of all memory cells: the total repertoire of subject matters the agent has grasped. To simplify the notation, we set $\bigcap \mathcal{O} := O^{\cap}$ and $\oplus (\bigcup_{O \in \mathcal{O}} t(O)) := \mathfrak{b}$. Then the LTM of the agent is $O^{\cap}_{\mathfrak{b}}$, which features the 'total topic' the agent is on top of. Notice that $\mathfrak{b}$ is guaranteed to be in $\mathcal{T}$, since $\bigcup_{O \in \mathcal{O}} t(O)$ is finite.

LTM is larger than any single memory cell which can be activated as, or put into, WM, with respect to both information and topic. The agent passively believes, i.e., has in LTM, way more than it can actively believe, i.e., activate and process in WM: the latter has quite limited capacity compared to LTM, as cognitive psychology has taught us.

Next, a *model* $\mathfrak{M} = \langle W, \mathcal{O}, \mathcal{T}, \oplus, t, \Vdash \rangle$ is a frame with an interpretation $\Vdash$, which works differently from what we've seen in previous chapters: we now evaluate formulas with respect to world-memory pairs, $\langle w, O_x \rangle$, with $w \in W$ representing the actual world, and O_x a memory cell. The working memory WM is just the designated world-memory cell with respect to which we evaluate formulas. We denote the set of all world-memory pairs of model $\mathfrak{M}$ as $\mathcal{P}(\mathfrak{M})$ ('$\mathcal{P}$' is for 'pair', not the power set operation). The interpretation relates such pairs to atomic formulas: we read '$\langle w, O_x \rangle \Vdash p$' as saying that p holds at $\langle w, O_x \rangle$, '$\langle w, O_x \rangle \nVdash p$' as: $\sim \langle w, O_x \rangle \Vdash p$. This is extended to all formulas of $\mathcal{L}$ thus:

(S¬) $\langle w, O_x \rangle \Vdash \neg \varphi \Leftrightarrow \langle w, O_x \rangle \nVdash \varphi$

(S∧) $\langle w, O_x \rangle \Vdash \varphi \wedge \psi \Leftrightarrow \langle w, O_x \rangle \Vdash \varphi$ & $\langle w, O_x \rangle \Vdash \psi$

(S∨) $\langle w, O_x \rangle \Vdash \varphi \vee \psi \Leftrightarrow \langle w, O_x \rangle \Vdash \varphi$ *or* $\langle w, O_x \rangle \Vdash \psi$

(S□) $\langle w, O_x \rangle \Vdash \Box \varphi \Leftrightarrow W \subseteq |\varphi|^{O_x}$

(SB$_A$) $\langle w, O_x \rangle \Vdash B_A \varphi \Leftrightarrow$ [1] $O \subseteq |\varphi|^{O_x}$ & [2] $t(\varphi) \leq x$

(SB$_P$) $\langle w, O_x \rangle \Vdash B_P \varphi \Leftrightarrow$ [1] $O^\cap \subseteq |\varphi|^{O_x}$ & [2] $t(\varphi) \leq \mathfrak{b}$

where $|\varphi|^{O_x} = \{w \in W | \langle w, O_x \rangle \Vdash \varphi\}$.

Both active and passive belief are topic-sensitive and get TSIM-style, two-component truth conditions. For $B_* \varphi$ to come out true at $\langle w, O_x \rangle$, we ask for two things to happen: [1] φ must be entailed by the information O in WM for active belief, and by the information $O^\cap$ in LTM for passive belief; and [2] the topic of φ must be included in the topic x activated in WM, for active belief, and in the overall LTM topic $\mathfrak{b}$ the agent is on top of, for passive belief.

Only the truth value of an ascription of *active* belief depends on the chosen O_x.[1] However, the agent can believe

---

[1] Given a model $\mathfrak{M} = \langle W, \mathcal{O}, \mathcal{T}, \oplus, t, \Vdash \rangle$, $w \in W$, two world-memory pairs $(w, O_x), (w, U_y) \in \mathcal{P}(\mathfrak{M})$, and $\varphi \in \mathcal{L}$ such that φ does not have any occurrences of B_A, we have: $\langle w, O_x \rangle \Vdash \varphi \Leftrightarrow \langle w, U_y \rangle \Vdash \varphi$.

φ with respect to one memory cell without believing the same content with respect to another one. That is, given a model $\mathfrak{M} = \langle W, \mathcal{O}, \mathcal{T}, \oplus, t, \Vdash \rangle$ and two world-memory pairs $\langle w, O_x \rangle, \langle w, U_y \rangle \in \mathcal{P}(\mathfrak{M})$, it could be that $\langle w, O_x \rangle \Vdash B_A \varphi$ and $\langle w, U_y \rangle \nVdash B_A \varphi$ for some $\varphi \in \mathcal{L}$, as shown in the Sample Model in the footnote.[2]

Finally, valid entailment is truth preservation at all world-memory pairs of all models. With Σ a set of formulas:

$$\Sigma \vDash \psi \Leftrightarrow \text{ in all models } \mathfrak{M} = \langle W, \mathcal{O}, \mathcal{T}, \oplus, t, \Vdash \rangle \text{ and for all}$$
$$\langle w, O_x \rangle \in \mathcal{P}(\mathfrak{M}): \langle w, O_x \rangle \Vdash \varphi \text{ for all } \varphi \in \Sigma \Rightarrow \langle w, O_x \rangle \Vdash \psi$$

For single-premise entailment, we write $\varphi \vDash \psi$ for $\{\varphi\} \vDash \psi$. Validity for formulas, $\vDash \varphi$, truth at all world-memory pairs of all models, is $\emptyset \vDash \varphi$, entailment by the empty set of premises.

We'll make use of the abbreviation $\bar{\varphi} := \bigwedge_{p \in \mathfrak{At}\varphi}(p \vee \neg p)$. This will play a role in formalizing validities and invalidities.[3]

---

[2]Let $\mathfrak{M} = \langle \{w, w_1, w_2\}, \{O, U\}, \{x, \mathfrak{b}, y, z\}, \oplus, t, \Vdash \rangle$ such that $O = \{w, w_1\}$, $U = \{w, w_2\}$, and $\langle \{x, \mathfrak{b}, y, z\}, \oplus \rangle$ constitutes the join-semilattice in this figure:

The dots are topics and the lines represent topic-inclusion relations, going upwards. So for t we have $t(p) = x$, $t(q) = y$, $t(r) = z$. As for $\Vdash$, p and q's truth set is $\{w, w_1\}$, r's truth set is $\{w, w_2\}$. Then $(w, O_x) \Vdash B_A p$ since $O \subseteq \{w, w_1\}$ and $t(p) \leq x$. However, $(w, O_y) \nVdash B_A p$ since $t(p) \nleq y$, that is, the agent does not have the subject matter of p in working memory O_y. Similarly, we also have, e.g., $(w, U_y) \nVdash B_A p$ for two reasons: (1) $U \nsubseteq \{w, w_1\}$ and (2) $t(p) \nleq y$, that is, the informational content of U_y does not eliminate all non-p possibilities and the subject matter of p is not part of the subject matter of working memory U_y, respectively.

[3]In order to have a unique definition of each $\bar{\varphi}$, we set the convention that elements of $\mathfrak{At}\varphi$ occur in $\bigwedge_{p \in \mathfrak{At}\varphi}(p \vee \neg p)$ from left-to-right in the order they are enumerated in $\mathcal{L}_{AT} = \{p_1, p_2, \dots\}$. For example, for $\varphi := B_*(p_{10} \to p_2) \vee \Box p_7$, $\bar{\varphi}$ is $(p_2 \vee \neg p_2) \wedge (p_7 \vee \neg p_7) \wedge (p_{10} \vee \neg p_{10})$, and not $(p_{10} \vee \neg p_{10}) \wedge (p_7 \vee \neg p_7) \wedge (p_2 \vee \neg p_2)$ or $(p_7 \vee \neg p_7) \wedge (p_{10} \vee \neg p_{10}) \wedge (p_2 \vee \neg p_2)$ etc. This convention will eventually not matter since our logic cannot differentiate two conjunctions of different order: $\varphi \wedge \psi$ is provably and semantically equivalent to $\psi \wedge \varphi$.

Given a model $\mathfrak{M} = \langle W, \mathcal{O}, \mathcal{T}, \oplus, t, \Vdash \rangle$, it is easy to see that $\bar{\varphi}$ is true at every world-memory pair in $\mathcal{P}(\mathfrak{M})$ and $\mathfrak{At}\bar{\varphi} = \mathfrak{At}\varphi$ for any $\varphi \in \mathcal{L}$. This trick allows us to talk inside the language about what topics the agent is actively thinking about in WM, and what topics the agent has grasped and stored in LTM. Formulas of the form $B_A\bar{\varphi}$ ($\neg B_A\bar{\varphi}$) express within $\mathcal{L}$ statements such as 'The agent has (does not have) the subject matter of φ in WM'.[4] Similarly, formulas of the form $B_P\bar{\varphi}$ ($\neg B_P\bar{\varphi}$) express within $\mathcal{L}$ statements such as 'The agent has (does not have) the subject matter of φ in LTM'.

Our semantics is a variation on the *subset space semantics* of Moss and Parikh (1992), in that the component $\langle W, \mathcal{O} \rangle$ of our frames is a subset space (a pair of a set and a selection of its subsets) and we evaluate sentences not at worlds but at world-set pairs. Subset space semantics was originally designed to model an evidence-based notion of absolutely certain knowledge and epistemic effort. The evaluation pairs of the form $\langle w, O \rangle$ within this framework obey the constraint that $w \in O$ (for knowledge is veridical) and are often called 'epistemic scenarios'. O represents the agent's current truthful evidence.

Our framework comes with a distinct formalism, however, and a different interpretation of a subset space model's components. We focus on belief rather than knowledge, so the evaluation pairs are tailored accordingly: as belief is not factive, a memory cell $\langle w, O_x \rangle$ does not have to meet the constraint $w \in O$. More importantly, our subset spaces and the corresponding evaluation pairs are endowed with topics. This makes the resulting logic of belief hyperintensional, as opposed to the intensional epistemic logics of the traditional subset space semantics (Moss and Parikh 1992; Dabrowski et al. 1996; Weiss and Parikh 2002).

7.5   The Logic of Framing

In Berto and Özgün (2021), we come up with a sound and complete axiomatization L for the logic of framed belief over

[4]Notice that $\langle w, O_x \rangle \Vdash B_A\bar{\varphi} \Leftrightarrow O \subseteq |\bar{\varphi}|^{O_x}$ & $t(\bar{\varphi}) \leq x \Leftrightarrow O \subseteq W$ & $t(\varphi) \leq x \Leftrightarrow t(\varphi) \leq x$.

$\mathcal{L}$. The axioms are:

(CPL) All classical tautologies and Modus Ponens;

(S5$_\square$) S5 axioms and rules for $\square$;

(I) Axioms for B_*, with $* \in \{A, P\}$:

(C_{B_*}) $B_*(\varphi \wedge \psi) \equiv (B_*\varphi \wedge B_*\psi)$

$(\mathsf{Ax1}_{B_*})$ $B_*\varphi \supset B_*\overline{\varphi}$

$(\mathsf{Ax2}_{B_*})$ $(\square(\varphi \supset \psi) \wedge B_*\varphi \wedge B_*\overline{\psi}) \supset B_*\psi$

$(\mathsf{Ax3}_{B_*})$ $B_*\varphi \supset \square B_*\varphi$

(II) Axioms for B_A:

(D_{B_A}) $B_A\varphi \supset \neg B_A\neg\varphi$

(III) Axioms connecting B_A and B_P:

(Inc) $B_A\varphi \supset B_P\varphi$

The notion of derivation, denoted by $\vdash$, in $\mathfrak{L}$ is defined as usual. Thus, $\vdash \varphi$ means φ is a theorem of $\mathfrak{L}$. $\mathfrak{L}$ is a sound and complete axiomatization of $\mathcal{L}$ with respect to the class of models given above: for every $\varphi \in \mathcal{L}$, $\vdash \varphi$ if and only if $\vDash \varphi$ (see the appendix to our paper for the proof).

The axioms in Group I give general closure features of belief, both active and passive, for our framed agents. C_{B_*} ensures that beliefs are fully Conjunctive, as usual for our TSIMs and as per the defence in Section 3.2: one who believes, either actively or passively, that John is tall and handsome, believes both that John is tall and that John is handsome, and vice versa. $\mathsf{Ax1}_{B_*}$ captures, as desired, the topic-sensitivity of belief: one can actively believe φ only if one is actively thinking about the relevant topic in WM; one can passively believe φ only if one has concepts for the relevant topic stored in LTM. $\mathsf{Ax2}_{B_*}$ states a limited deductive closure principle for both active and passive belief: if ψ follows from φ a priori, and one believes φ, and one is on top of the subject matter of ψ, then one does believe ψ. $\mathsf{Ax3}_{B_*}$ has it that beliefs are not world-relative.

In Group II, D_{B_A} states a consistency principle for active belief: one who has φ in WM will not also have $\neg\varphi$ there. Notice that this does not hold for passive belief: our framed agent may have all sorts of inconsistent beliefs stored or encoded in its LTM. They can stay there insofar as one does not think about them all together. This makes for a very realistic modelling: isn't this the way we are, for the most part? We are quite inconsistent in the beliefs we hold – provided the inconsistencies remain stored in our long-term memory, shielded from the focus of our attention.

As for Group III, the Inc principle bridges active and passive belief. It guarantees, as desired, that whatever is activated in WM be available in LTM to begin with.

As always with our TSIMs, just as important as validities are the invalidities involving them, as they display the precise sort of non-omniscience our framed agents instantiate. We discuss a few prominent invalidities:

1. From φ, infer $B_*\varphi$ [Omniscience Rule]

2. $\Box\varphi \supset B_*\varphi$ [A Priori Omniscience]

3. $(\Box(\psi \supset \psi) \wedge B_*\varphi) \supset B_*\psi$ [Closure under A Priori Implication]

4. $\neg B_*\varphi \supset B_*\neg B_*\varphi$ [Negative Introspection]

5. From $\varphi \equiv \psi$, infer $B_*\varphi \equiv B_*\psi$ [Framing-A]

6. $(B_A\varphi \wedge B_P(\varphi \equiv \psi)) \supset B_A\psi$ [Framing-B]

7. From $\varphi \equiv \psi$, infer $(B_A\varphi \wedge B_P\overline{\psi}) \supset B_A\psi$ [Framing-C][5]

---

[5] *Countermodel*: take our Sample Model from footnote 2 above. We have (1) and (2) invalid since $\vDash r \vee \neg r$ (therefore also $(w, O_x) \Vdash \Box(r \vee \neg r)$), but $(w, O_x) \nVdash B_A(r \vee \neg r)$ (since $t(r) \not\leq x$) and $(w, O_x) \nVdash B_P(r \vee \neg r)$ (since $t(r) \not\leq b$). For (3), take $\varphi := p$ and $\psi := r \vee \neg r$: $(w, O_x) \Vdash \Box(p \to (r \vee \neg r))$, $(w, O_x) \Vdash B_A p$, and $(w, O_x) \Vdash B_P p$, however, $(w, O_x) \nVdash B_A(r \vee \neg r)$, and $(w, O_x) \nVdash B_P(r \vee \neg r)$ as shown above. For (4), take $\varphi := r$: world-memory pair (w, O_x) falsifies it for B_A since $t(r) = t(\neg B_A r) \not\leq x$ and falsifies it for B_P since $t(r) = t(\neg B_P r) \not\leq b$. For (5), take $\varphi := p \vee \neg p$ and $\psi := r \vee \neg r$, and (w, O_x) falsifies the principle. For (6), take $\varphi := p$ and $\psi := q$: $(w, O_x) \Vdash B_A p$ and $(w, O_x) \Vdash B_P(p \leftrightarrow q)$, but $(w, O_x) \nVdash B_A q$ (since $t(q) \not\leq x$). For (7), take $\varphi := p \vee \neg p$ and $\psi := q \vee \neg q$, and observe that (w, O_x) falsifies the principle.

The failure of (1)-(3) tells us that our agents don't believe all (*a priori*) truths and that their beliefs are not closed under *a priori* implication. (4) says that they lack the wisdom of negative introspection: they can fail to believe that they don't believe something.

The last three invalidities, (5)-(7), crucially capture the typical framing we were after: Framing-A guarantees that agents can have different attitudes towards intensionally equivalent formulas. Framing-B says that one can have the belief that φ (e.g., patients should get surgery with a 90% one-month survival rate) activated in WM, without having the belief that ψ (patients should get surgery with a 10% first-month mortality) there, even when one *does* have their equivalence in one's LTM. In this sense, one is aware: one is on top of all the relevant concepts and does believe that either is true iff the other is. But all of this is left asleep in LTM. In this other sense, one is not aware: one is just not thinking about it. Framing-C says that one's actively believing φ does not imply that one actively believes ψ, even when the two are equivalent and one has the subject matter of ψ in one's LTM.

Here's something the logic does not capture (an admission prompted by a remark by one reviewer of this book): the positive/negative *polarity* displayed by pairs of claims involved in typical cases of framing – death and survival rates, penalties and discounts, etc. The reviewer graciously granted that perhaps a logic of framing is not supposed to do this in a general setting. We hope so, for right now, we don't know how to tweak ours so that it does.

We close the chapter by mentioning two directions of further investigation: first, both active and passive belief TSIMs are plain, categorical forms of belief. It may be interesting to expand the language and formal semantics so that they include conditional, topic-sensitive active and passive belief, as per the two-place TSIMs we explored in previous chapters.

Second, working memory is properly so-called in cognitive psychology because it is the locus of cognitive activity: beliefs are in there in order to be manipulated, expanded, revised via operations of combination, deduction, etc. Another direction of expansion may then feature the addition to our language of topic-sensitive dynamic operators in the style of Dynamic Epistemic Logic, perhaps as per the route summarized in Section 6.4. This would allow one to properly model how agents operate on their active beliefs in the light of new incoming information, before storing the results in LTM.

7.6   Chapter Summary

This chapter has introduced two kinds of one-place TSIMs representing, respectively, belief activated in working memory, and belief left passively stored in long-term memory. The distinction between the two sorts of belief has been shown to model a typical form of the well-known framing effect, whereby people can have different attitudes towards logically or necessarily equivalent propositions. The chapter has introduced a semantics for active and passive topic-sensitive belief to represent, and reason about, agents whose belief states can be subject to framing effects. The analysis of framing has called for a precise characterization of the sense in which framed agents are logically non-omniscient, given that they can believe exactly one of two intensionally equivalent propositions even when they are fully on top of the relevant subject matters and, in a 'dormant' sense, they are aware of the equivalence.

8

Probabilities, Indicatives, and Relevance

Co-authored with Aybüke Özgün

In this final chapter, we combine topic-sensitivity and probabilities in the abstract and general setting of conditionality. We start from *Adams' Thesis*, which is often taken as capturing a fundamental connection between probabilities and conditionals: it claims that the acceptability of a simple indicative equals the corresponding conditional probability. (A simple indicative $\varphi \to \psi$ is one with no indicatives in φ or in ψ; examples of non-simple indicatives: 'If you pass the exam, then if Mary passes, too, you'll have a party'; 'If John goes to the concert if Mary goes, then John has changed his mind'.) The Thesis is widely endorsed by philosophers but, as we show in Section 8.1, it is arguably false and refuted by empirical research on how and when people accept indicatives.

To fix Adams' Thesis, we submit, we need a *relevance* constraint: we accept a simple indicative conditional $\varphi \to \psi$ to the extent that (i) the conditional probability $p(\psi|\varphi)$ is high, provided that (ii) φ is relevant for ψ. How (i) should work is well-understood. It is (ii) that holds the key to advance our understanding of conditionals. You may not be surprised to hear that such an advancement can come, we claim, by taking relevance as topic-sensitivity. As we also show in that section, relevance does not easily reduce to Gricean pragmatics.

Some approaches to relevance for indicatives, in particular the *inferentialist* and *evidential support* views, are discussed

Topics of Thought: The Logic of Knowledge, Belief, Imagination.
Francesco Berto, Oxford University Press. © Francesco Berto 2022.
DOI: 10.1093/oso/9780192857491.003.0008

in Section 8.2, paving the way to our proposal, which comes in Section 8.3. Here, we stay neutral on the issue of whether indicatives have truth conditions at all, whereas we propose a formal framework giving acceptability and logical closure conditions for simple indicatives: its probabilistic component (i) uses Popper functions; its relevance component (ii) is given via our familiar algebraic structure of topics.

In Section 8.4, we then describe the resulting probabilistic logic, reporting some technical results presented in more detail in our Berto and Özgün (2021). We argue that its (in)validities are not just theoretically desirable, but also in line with empirical studies, whose results we briefly summarize there, on how people reason with conditionals.

8.1   Adams' Thesis and the Problem of Relevance

Adams' Thesis (Adams 1966, 1975) has it that the *acceptability* of a simple indicative conditional $\varphi \to \psi$ equals the corresponding conditional probability:

(AT) $Acc(\varphi \to \psi) = p(\psi|\varphi)$

AT is sometimes, but should not be, confused with *Stalnaker's Hypothesis* (Stalnaker 1975), also called 'the Equation' by authors like Edgington and Bennett, and which has it that the *probability* of an indicative $\varphi \to \psi$ equals the corresponding conditional probability:

(SH) $p(\varphi \to \psi) = p(\psi|\varphi)$

Conditional probabilities are standardly understood as ratios of unconditional ones: $p(\psi|\varphi) = \frac{p(\psi \wedge \varphi)}{p(\varphi)}$ when $p(\varphi) > 0$; and undefined otherwise. AT and SH are at times formulated with the proviso that $p(\psi|\varphi) = 1$ when $p(\varphi) = 0$ for the conditional probability to be defined for all φ (see, e.g., Adams 1998, 150).

AT is popular in philosophy among proponents of the non-propositional view of indicatives (Edgington 1995; Adams 1998; Bennett 2003). SH is popular in psychology: it is in line with the New Paradigm Psychology of Reasoning (Over 2009; Elqayam and Over 2013), which puts probabilities at centre stage in the study of reasoning, and handles conditionals probabilistically (Evans and Over 2004; Oaksford and Chater 2010). The Paradigm is becoming so dominant that even proponents of essentially non-probabilistic accounts of the conditional, like the mental models theory (Johnson-Laird and Byrne 2002), feel the need to relate their view to probabilities (Girotto and Johnson-Laird 2010).

However, Lewis' and others' notorious triviality results (Lewis 1976; Hajek 1989) are often taken as showing that SH can't be *quite* right. On the other hand, by endorsing AT rather than SH, non-propositionalists can insist that indicatives be handled probabilistically: they are safe from triviality for they don't express propositions and cannot generally be embedded, hence the limitation to simple conditionals in AT. They generally lack truth values, or they lack a complete truth table (one may take them as false when the antecedent is true and the consequent false, true when both are true: see e.g. the 'ersatz truth values' of Adams (1998), 121-3; Bennett (2003), ch. 8. Thus, they lack probabilities of truth properly so called, as Adams realized, and so it might be misleading to say that they have believability conditions if to believe φ is to believe that φ is true. But they can have acceptability conditions, as per AT.

What is acceptability once it's disentangled from believability, a reviewer of this book asks? Good question – Douven (2016), ch. 4, has a nice discussion. One may say that if $\varphi \to \psi$ doesn't express a proposition and lacks a truth value, accepting it is, or at least requires, being disposed to do certain things with it, which align to what one is disposed to do when one believes something that does have a truth value. E.g., if one accepts $p \to q$ and one believes that p, one is prone to believe q as well: see Bennett (2003), ch. 8. (Whether one *should* believe q in general is a complicated issue, as Harman (1986) has taught us.)

Several philosophers like AT, and in particular, as shown by Douven (2016), ch. 4, they often take it as a principle which describes what people generally do, not as a normative principle. McGee (1986) claims that '[AT] describes what

English speakers assert and accept with unfailing accuracy'
(485). And Jackson says:

> There is a great deal of evidence for [AT]. There
> is head-counting evidence. Very many philosophers
> of otherwise differing opinions have found [AT]
> highly intuitive. There is case-by-case evidence.
> Take a conditional which is highly assertible [...]; it
> will invariably be one whose consequent is highly
> probable given its antecedent. (Jackson 1987, 12)

(Notice that McGee and Jackson speak – also – about assertion. We'll come back to how this connects to acceptance.)

AT is false. A conditional probability $p(\psi|\varphi)$ for an unac-
ceptable indicative can be high because ψ is already likely
and has little to do with φ:

1. If Brexit causes a recession, then Jupiter is a planet.

One may claim that (1) is unacceptable for its consequent has
probability 1. We'll come back to the issue of conditionals
with extreme antecedent or consequent probabilities. Even
granting the claim, sometimes conditionals with high but
less than 1 probability of their consequent are unacceptable
because this has little to do with their antecedent:

2. If Brexit causes a recession, then there will be some
   heads in the first hundred tosses of this fair coin.

AT is empirically inadequate, too. In the experiments
reported by Douven and Verbrugge (2010), one group of
subjects was given contexts C_i, $1 \leq i \leq 30$, and asked to
rate the acceptability of conditionals $\varphi_i \to \psi_i$ in C_i. Another
group was given the same contexts C_i and asked to judge the
probability of ψ_i in C_i on the supposition that φ_i. People's
patterns (of degrees) of acceptance for conditionals generally
don't even approximate the corresponding conditional probabilities: this 'manifestly refute[s] Adams' Thesis, both in its
strict form AT and in its approximate form' (Douven 2016,
99).

One should not confuse the empirical support for SH with
the (lack of) empirical support for AT. As noted also in

Douven and Verbrugge (2010), sect. 4, there is significant experimental work that supports SH, finding high correlation between the probabilities that the participants assign to conditionals and the corresponding conditional probabilities. However, to the best of our knowledge, Douven and Verbrugge are the first to test AT by asking a group of participants to grade the acceptability of conditionals rather than their probability of truth. We refer to the aforementioned source for further references of empirical results supporting SH and a detailed discussion on how experiments on AT and SH differ.

The conditionals that fare better with respect to AT are what Douven and Verbrugge call 'deductive inferential': these are conditionals such that their consequent follows deductively from the antecedent plus background, unstated assumptions; for these, at least, a high correlation was found.

What's wrong with AT? Compare (2) above with the following, adapting Douven (2016), 104:

3. If there's some heads in the first ten tosses, then there will be some heads in the first hundred tosses of this fair coin.

We accept (3), not (2), because (3)'s antecedent is relevant for the consequent, which is, instead, off-topic with respect to (2)'s antecedent. This suggests a fixing for AT: we accept a conditional to the extent that (i) the consequent is likely conditional on the antecedent, *provided* (ii) some relevance connection linking antecedent and consequent is satisfied.

What is relevance for indicatives? A venerable idea going back to Grice (1989), and sometimes invoked to save the material conditional analysis from apparent counterexamples, is that relevance is a pragmatic issue: some perfectly true or probabilistically all right conditionals are unassertable, lacking a connection between antecedent and consequent (Johnson-Laird and Byrne 1991).

What is the connection between assertability and acceptability? We take acceptance as a mental state, assertion as a linguistic act manifesting acceptance (or belief), when one speaks sincerely. We speak of assertability in strictly pragmatic contexts, but we are after acceptability conditions: we follow Douven (2016), 94, in taking the latter as the core notion. The Douven and Verbrugge (2010) experiments were explicitly designed and phrased in terms of acceptability, which is not subject to social norms the way assertability is: one may find something very acceptable and reasonable, but inappropriate to assert in a given conversational context, e.g., because it would be considered weird, or an insensitive thing to say, or so.

We shouldn't take for granted that relevance for indicatives has to be handled in a broadly Gricean way, as merely involving cancellable pragmatic implicatures. Sophisticated approaches to the logic of conditionals, such as relevance logics (Dunn and Restall 2002), make relevance amenable to a rigorous, compositional, and properly semantic treatment, and have also been developed for *ceteris paribus* conditionals (Mares and Fuhrmann 1995; Mares 2004).

Besides, as Krzyżanowska et al. (2017) have shown, pragmatic coherence concerning what is and is not assertable may be a weaker constraint than proper relevance of conditional antecedents for their consequents: the former typically requires, e.g., that one not assert a conjunction when the two conjuncts have nothing to do with each other ('Brexit will cause a recession and Jupiter is a planet' is an odd thing to say in one breath, more or less in any natural conversational context). However, sometimes a conjunction is assertable because of some topic overlap between the two conjuncts, but we don't want to assert the corresponding conditional. Picking Krzyżanowska et al.'s own example, one can easily think of contexts where this is clearly assertable:

4. Raccoons have no wings and they cannot breathe under water.

What makes (4) pragmatically all right is that the two conjuncts overlap in topic: both are about raccoons and what they are like. But we may not want to assert, in the very same contexts in which (4) is a fine thing to say, the corresponding conditional:

5. If raccoons have no wings, then they cannot breathe under water.

Or take this (now drawing on (Priest 2008, 96)):

6. A fortune-teller predicts that you'll win the lottery, and you do.

It's pragmatically all right to assert (6) in a number of contexts, for its conjuncts overlap in topic: both have to do with your winning the lottery. But we may not want to assert, in those same contexts, the corresponding superstitious conditional:

7. If a fortune-teller predicts that you'll win the lottery, then you do.

Krzyżanowska et al. (2017) exhibit experimental results giving some evidence that, even when people find it pragmatically appropriate to assert conjunctions like (4) or (6), they tend not to assert the corresponding irrelevant conditionals, like (5) and (7).

The moves from (4) to (5), or from (6) to (7), present instances of the And-to-If schema, licensing the inference from a conjunction to the corresponding conditional:

(And-to-If) $\varphi \wedge \psi \vDash \varphi \rightarrow \psi$

Entailment, $\vDash$, here may be understood standardly, as truth preservation (in all models), or, if indicatives lack truth conditions, as preservation of degrees of probability, or of acceptability, or so, as e.g. in Adams (1998). And-to-If is sometimes called 'Centering', for it holds in the similarity-based possible worlds semantics for conditionals due to Stalnaker (1968) and Lewis (1973), when one assumes that the world of evaluation is always the single world most similar to itself (it's the unique one at the centre of the nested spheres of worlds arranged around it). It doesn't hold only there. A number of mainstream theories of indicatives validate And-to-If: the material conditional view (Jackson 1987; Grice 1989) and the probabilistic-suppositional view (Adams 1975; Edgington 1995; Evans and Over 2004), for instance, have it.

At least insofar as *acceptability* is concerned, we think, And-to-If does not work: even if it turns out to be necessarily truth-preserving and/or probabilistically valid, it's not acceptability-preserving. An acceptable conjunction doesn't generally warrant the acceptability of the corresponding conditional. And the mismatch between the two is not easily reducible to Gricean pragmatics and cancellable implicatures, even if it turns out not to be properly addressed by making And-to-If fail in a truth-conditional or probabilistic account of entailment. In a recent and thoroughly argued paper, Dan Lassiter (2021), for instance, spells trouble for views such as the ones to be explored in the following Section 8.2, insofar as they embed relevance in the truth conditions for indicatives. According to Lassiter, relevance has to do with coherence constraints linking subsequent clauses in discourse and should not be embedded in the semantic content of indicatives. But even for him, the coherence relations which are to account for relevance are a *mandatory* part of the interpretation of discourse: they cannot be cancelled the way pragmatic implicatures normally can. Meanwhile, exploring a couple of views that make And-to-If fail will allow a number of useful considerations to emerge, in view of the presentation of our own account starting in Section 8.3.

8.2   Inferentialism and Evidential Support

Some *inferentialist* (Braine 1978; Braine and O'Brien 1991) approaches to conditionals have it that conditionals express enthymematic arguments. Actually, the label 'inferentialism' is used more generally in the literature, to refer more or less to any account that emphasizes relevance as influencing the truth or acceptability conditions of conditionals. So used, the label would apply to the evidential support theory to be discussed below, to other approaches that handle relevance probabilistically, e.g., Skovgaard-Olsen et al. (2016), or causally e.g., Van Rooij and Schulz (2019), or by resorting to non-classical logics, e.g., Dunn and Restall (2002), and to our own view as well. But the label, however popular, is a bit of a misnomer: as we will see, relevance needn't be understood as inferential, unless one stretches 'inferential' beyond usefulness.

As for inferentialism properly so named, the idea goes back to Mill's *System of Logic*. It was endorsed by Ramsey in the same work where he introduced the 'test' we mentioned in

Section 5.1, whereby we evaluate a conditional by supposing the antecedent and assessing the consequent under that supposition:

> [W]e can say with Mill that 'If p, then q' means that q is inferrible from p, that is, of course, from p together with certain facts and laws not stated but in some way indicated by the context. (Ramsey 1990, 156)

So $\varphi \to \psi$ says that there's some good inference from φ and background assumptions ('facts and laws') BA_φ to ψ. Besides plausibly depending on the antecedent (that's what the subscript is there for), background assumptions depend on context and their list can be open-ended. They capture the idea that everyday conditionals are for the most part *ceteris paribus* and non-monotonic.

For lots of good conditionals, there is no way to deduce ψ from φ, no matter what BA_φ come to help. But we needn't assume that the valid inference at issue be deductive: ψ may follow from φ and BA_φ in other ways too: inductively, abductively, or via a mixture of different ways of inferring. Krzyżanowska (2015) imposes constraints on the connection between premises and conclusion which ensure that ψ doesn't follow trivially from φ and BA_φ, thus capturing a kind of relevance. The view makes And-to-If fail in a most natural way: the mere fact that φ and ψ are true together doesn't warrant there being a good argument from the former (and, BA_φ) to the latter.

Inferentialism (of this kind) has not been proposed, as far as we know, as a general account of indicatives. It can hardly be one. Linguists distinguish inferential from content conditionals (Declerck and Reed 2001; Haegeman 2003; Dancygier and Sweetser 2005) expressing non-logical connections between states of affairs: 'If John passes the exam, we'll have a party'; 'She's such a disappointment if she thinks so highly of him'. As stressed by Douven, the connections between antecedent and consequent in relevant conditionals can be of the most diverse kinds:

> [C]onditionals have been said to require for their truth the presence of a 'connection' linking their

> antecedent and consequent. Proposals in this vein immediately raise the question of what the nature of the supposed connection could be. Candidate answers abound: it could be logical, statistical, causal, explanatory, metaphysical, epistemic; or the 'connector' could be a second-order functional property, notably, the property that there is some first-order property or other that links antecedent and consequent. (Douven 2016, 35-6)

It's dubious that all relevant conditionals express the existence of some argument from their antecedent and contextually determined background assumptions to their consequent. It is surely in agreement with the Ramsey test to say that their assessment always involves some form of mental simulation, whereby we assess the consequent under the supposition of the antecedent. To label the process 'inferential' in all cases just on this basis, however, would be to stretch the term beyond usefulness: surely any conditional trivially says that its consequent follows, in some sense or other, from its antecedent! A general account of indicatives calls for a general notion of relevance. By resorting to topics, the criterion of relevance proposed in our account below aims at giving a catch-all condition, covering relevance of *any* kind, whether inferential or not.

Next, whenever an argument condensed in $\varphi \to \psi$ is not purely deductive, it may be valid even when φ is true and ψ isn't: good arguments involving inductive or abductive steps may fail to be necessarily truth-preserving. Thus, inferentialism is bound to invalidate Modus Ponens (Krzyżanowska 2015, 70-1). But preserving X forwards (X being truth, or degrees of probability, of acceptability, or whatnot) has often been taken as a minimal requirement for an operator to count as a conditional. Putative exceptions are very controversial, and anyway involve peculiar sentences (paradoxes like the Liar, see Beall (2015)), or contexts like the famous McGee cases – which anyway don't affect *simple* conditionals, as they crucially involve right-nested ones. As McGee himself admitted, 'there is every reason to suppose that, restricted to [simple] conditionals, modus ponens is unexceptionable' (McGee 1985, 468). Additionally, with over 97% endorsement across a range of empirical tests (Oaksford 2005; Oaksford and Chater 2010), Modus Ponens is a very popular inference involving conditionals, and one of the most popular *tout-court* (Evans and Over 2004, 46–52).

The *evidential support thesis* (EST) championed by Douven (2016) proposes to fix AT by adding to it a relevance condition of evidential support. Evidence is understood probabilistically: φ is evidence for ψ by making it more likely. The qualitative (non-graded) acceptability conditions for a simple indicative are:

(EST) $Acc(\varphi \rightarrow \psi)$ iff (i) $p(\psi|\varphi) > \theta$ and (ii) $p(\psi|\varphi) > p(\psi)$

(i) is a qualitative variant of AT, saying that the conditional probability passes a threshold (say, $\theta \in [0.5, 1)$). (ii) is the evidential constraint: ψ is more likely conditional on φ than it is unconditionally. And-to-If nicely fails: $\varphi \wedge \psi$ can be true and acceptable without φ raising one bit the probability of ψ. This seems to be going on in a number of cases where we don't accept a conditional with true antecedent and consequent.

One issue with the view is its inferential weakness. Douven defines a notion of entailment as acceptability-preservation: when all premises reach a threshold of acceptability θ, the conclusion does, too. An inference is *valid for t* when it's acceptability-preserving for $\theta = t$, *invalid for t* otherwise. An inference is *valid* (*invalid*) *simpliciter* when valid (invalid) for all $t \in [0.5, 1)$ (Douven 2016, 130). When '⊨' is such validity, we have the following failures for the EST conditional (we now name the (in)validities we discuss with the labels used by Douven himself, and taken from the literature on conditional logics):

(Modus Ponens) $\varphi \rightarrow \psi, \varphi \nvDash \psi$

(CC) $\varphi \rightarrow \psi, \varphi \rightarrow \chi \nvDash \varphi \rightarrow (\psi \wedge \chi)$ [Conjunction in the Consequent]

(CMon) $\varphi \rightarrow \psi, \varphi \rightarrow \chi \nvDash (\varphi \wedge \psi) \rightarrow \chi$ [Cautious Monotonicity]

(CT) $\varphi \rightarrow \psi, (\varphi \wedge \psi) \rightarrow \chi \nvDash \varphi \rightarrow \chi$ [Cautious Transitivity]

We've already highlighted the badness of Modus Ponens failure. Segerberg (1989) claims that CC should hold in *any* reasonable system of conditional logic. CC, however, is the counterpart of what we have been calling Adjunction in various discussions concerning our two-place, all-or-nothing conditional-like TSIMs. We already mentioned in Section 3.2 that some intuitive pull against Adjunction can come from the consideration of probabilistic or degree-theoretic notions. Take the Lottery Paradox (Kyburg 1961): for each ticket i, $1 \leq i \leq n$, of a large enough fair lottery L, 'If L has exactly one winner then ticket i will lose' sounds acceptable, but 'If L has exactly one winner then ticket 1 will lose, and ticket 2 will lose, and ..., and ticket n will lose' doesn't. We'll come back to this in Section 8.4, after we have presented the logic of our own proposal: we will then see how it handles Lottery Paradox cases.

Meanwhile, there is wide agreement on CMon and CT, too, being required (they are, of course, just the counterparts of the Cautious Monotonicity and Cautious Transitivity principles, variously discussed for our two-place TSIMs). They feature in Chellas (1975)'s basic conditional logic and are valid also in our TSIM hyperintensional conditional belief setting of chapter 6. We've seen since Section 4.6 that Gabbay (1985) put them among the minimal requirements for a non-monotonic notion of entailment, and we mentioned in Section 6.2 that they hold in the system C of Kraus et al. (1990). Their popular non-monotonic logic P has them, too. It has been claimed that such principles are both theoretically and empirically desirable specifically for the indicative conditional (Pfeifer and Kleiter 2010): they are strong enough to do the job of the unrestricted Monotonicity (from $\varphi \to \psi$ to $\varphi \land \chi \to \psi$) and Transitivity (from $\varphi \to \psi$ and $\psi \to \chi$ to $\varphi \to \chi$) principles, which are often taken as invalid for *ceteris paribus* conditionals, while helping to explain why people sometimes endorse the latter by over-generalizing (Adams 1975; Bennett 2003; Pfeifer and Kleiter 2010).

The inferential weakness of EST may be tied to the particular way in which Douven defines his probabilistic logic: e.g., the recent Crupi and Iacona (2021) has a reworking of the evidential idea with an eye on this. But one issue besets the view due to its understanding relevance as probabilistic evidence: EST does not fare well with extreme probabilities.

If $p(\varphi) = 0$, φ can hardly be evidence for anything. If $p(\psi) = 1$, nothing can raise the probability of ψ. Then any conditional $\varphi \to \psi$ with 0 antecedent-probability or 1 consequent-probability is unacceptable. Many such conditionals, however (the relevant ones!), sound acceptable in a number of contexts. Douven (2016), 113, discusses one example of relevant conditional whose consequent has probability 1:

8. If Obama is president of the United States, his residence is in the White House.

While he finds (8) odd, we can think of a number of contexts in which it would be acceptable. In a plot to blackmail the president, the conspirators are pondering the best strategy. Suddenly one asserts: 'But if Obama is president, then his residence is in the White House; so we should infiltrate someone in the personnel working at the White House, who will manage to spy him; it's hard but not unfeasible'. If this can be done for a conditional like (8), whose antecedent and consequent are both not only true (at the time of Douven's writing), but also widely shared knowledge, it shouldn't be too difficult to find such contexts for a number of probability 1 consequents.

We don't want to insist too much on this, anyway, for we find the case of 0 probability antecedents more telling. *Pace* Bennett (2003) and others, one can non-trivially assess, and reason with, indicatives whose antecedent one fully takes to be false. One is pretty sure that Oswald killed Kennedy but has no troubles assessing 'If Oswald did not kill Kennedy, then someone else did' (Gillies 2004). As stressed by Joyce (1999), unpretentious thinkers can suppose *in the indicative mood* that φ also when they utterly disbelieve φ, and assess whether ψ is the case under that supposition:

[I]t is often assumed that any form of probabilistic belief revision that involves 'raising the dead' by increasing the probabilities of certainly false propositions must involve counterfactual beliefs. This is not so. It is logically consistent both to be certain that some proposition is false and yet to speculate about what the world is like if one is in fact wrong. To be subjectively certain of something is, after all, not the same as regarding oneself infallible on the matter. (Joyce 1999, 203)

This holds even for conditionals whose antecedents are taken as necessarily false:

9. If all even numbers are prime and 5 is even, then 5 is prime.

10. If all even numbers are prime and 5 is even, then one cannot square the circle.

(9) seems acceptable although its antecedent is a necessary falsity. Its same-antecedent (10) doesn't look acceptable – because a relevant connection with the consequent is missing. Douven mentions that these cases could be handled by resorting to a non-standard probabilistic account that doesn't assign probability 0 to all logical and mathematical falsehoods (Douven 2016, 114). The issue with (0), though, is that it seems to be acceptable also for one who is certain that its antecedent is false.

Cases like (9) and (10) give some evidence for a point at times though not always: see Jackson (1979); Bennett (2003) – neglected in the literature: the acceptability conditions for conditionals are hyperintensional. Just as our propositional attitudes in general are hyperintensional, as we have abundantly seen throughout this book, so can we sometimes have different attitudes towards conditionals (whether or not they will turn out to express propositions) whose antecedents and, respectively, consequents, are necessarily equivalent, having the same truth value across all possible worlds: we accept the relevant ones, not the irrelevant ones. Our account below will make conditionals hyperintensional precisely in this way, once again thanks to topic-sensitivity.

The so-called Ratio Formula, which defines a conditional probability $p(\psi|\varphi)$ as the ratio of two unconditional probabilities $\frac{p(\varphi \wedge \psi)}{p(\varphi)}$, makes a conditional probability undefined for $p(\varphi) = 0$. This should be taken as a problem for a treatment of conditionals that resorts to it, rather than for the claim that one can non-trivially reason with, or assess, indicatives with zero antecedent-probability. The use of Popper functions, whereby one doesn't define conditional probabilities via unconditional ones, is often recommended because they easily handle such cases. Several approaches to conditional belief and belief revision (Van Fraassen 1995; Arlo-Costa and Parikh 2005; Baltag and Smets 2008a), thus, endorse an extension of classical probability theory using Popper functions. We are doing the same for our account, to which we finally turn.[1]

8.3   Topic-Sensitive Probabilistic Semantics

We need a relevance constraint to fix AT: we accept $\varphi \to \psi$ to the extent that (i) $p(\psi|\varphi)$ is high, provided (ii) φ is relevant for ψ. Unlike EST, we understand relevance as topic-sensitivity: a relevant conditional is one whose consequent is *about* the right topic, as contextually determined by its antecedent. We focus on simple indicatives and give only graded acceptability conditions for them, not truth conditions, to accommodate non-propositionalist views.

We want to be able to conditionalize on 0 probabilities in a non-trivial way. We therefore use Popper functions, following

---

[1] Three more approaches to relevance for indicatives, which we won't discuss in detail, are Rott (2019); Van Rooij and Schulz (2019); and the influential Skovgaard-Olsen et al. (2016). We just mention that the Van Rooij-Schulz paper is based on the promising idea that relevance can be accounted for via a condition of dependence between antecedent and consequent understood as causal correlation. Van Rooij and Schulz argue that this is compatible with a general probabilistic view, insofar as it reduces to conditional probability in natural cases. As for Skovgaard-Olsen et al., it is based on the idea that the acceptability of a conditional (they actually phrase the result in terms of probability assignments) correlates well with the corresponding conditional probability precisely when their relevance condition is satisfied. They propose what they call the *Default and Penalty Hypothesis* (DPH): by default, people evaluate $\varphi \to \psi$ expecting the consequent to be positively relevant for the antecedent. When the expectation is fulfilled, they go for $p(\psi|\varphi)$. When not, people add a 'penalty' to their estimate. This is in line with our proposal below, where, as we will see, acceptability equals conditional probability when our relevance condition is satisfied, and drops otherwise.

Hawthorne (1996) and Leitgeb (2012), among others. We interpret the conditional probability measures subjectively-epistemically, not as objective frequencies, following the mainstream on indicatives (Adams 1966, 1975, 1998; McGee 1986; Douven 2016).

The degree of acceptability of a simple indicative $\varphi \to \psi$ is given, (i) as per AT, by the corresponding conditional probability, $P(\psi|\varphi)$ (we use capital 'P' now, to mark a setting in which conditional probabilities are taken as primitive), provided (ii) the conditional is *on-topic* – otherwise, $\varphi \to \psi$ has zero acceptability.

A conditional is on-topic when the topic of its consequent is fully included in a topic contextually determined by its antecedent. One may not take this latter as just the topic of the antecedent φ. Rather, one may take it as the topic of the relevant background assumptions BA_φ determined by φ and context (where, plausibly, $\varphi \in BA_\varphi$). That's because we sometimes accept $\varphi \to \psi$ in contexts where, given at least some intuitive way of understanding topicality, there seems to be no direct and plain topic-inclusion between φ and ψ:

11. If we keep burning fossil fuels at this pace, the polar ice will melt.

12. If Brexit causes a recession, the Tories won't win the next election.

13. If you push the button, the engine will start.

In cases like (11)-(13), the antecedent is relevant for the consequent although it doesn't, on its own, address an issue with respect to which the consequent is fully on-topic. Rather, the supposition of the antecedent triggers, in context, background assumptions with respect to which the consequent is fully on-topic (e.g., for (11), the topic of fossil fuel burning triggers topics such as those of the emission of CO_2, raising global temperatures, etc.). The topicality is between the background BA_φ and ψ – see Khoo (2016) for a recent view in the same ballpark. According to Khoo, what an indicative expresses is given by a contextually salient question under discussion, determining a partition of modal space. The link with topics should be patent.) The topic of BA_φ is determined, given that of φ, by a function f obeying plausible constraints.

Our base language $\mathcal{L}$ for this chapter will have the usual countable set $\mathcal{L}_{AT}$ of atomic formulas p, q, r $(p_1, p_2...)$, and, besides them, only negation $\neg$, conjunction $\wedge$, disjunction $\vee$. The well-formed formulas are the items in $\mathcal{L}_{AT}$ and, if φ and ψ are formulas, then so are:

$$\neg\varphi \mid (\varphi \wedge \psi) \mid (\varphi \vee \psi)$$

As usual, we often omit outermost brackets and we identify $\mathcal{L}$ with the set of its well-formed formulas, which we call *Boolean sentences*. It will come in handy to have a $\top := p \vee \neg p$ once more (and again, as in chapter 7, this is not to be confused with the $\top$ of Section 6.4: in particular, total subject matters don't matter here!) and a $\bot := \neg\top$. And once more, '$\mathfrak{At}\varphi$' stands for the set of atomic formulas occurring in φ. We use the notation '$\models_{PL}$' for logical truth/consequence as in classical propositional logic.

The full language $\mathcal{L}^{\rightarrow}$ of simple indicative conditionals expands $\mathcal{L}$ by an indicative conditional operator, $\rightarrow$, which connects only the elements of $\mathcal{L}$, so as to avoid nested conditionals: the well-formed formulas of $\mathcal{L}^{\rightarrow}$ are those of $\mathcal{L}$, plus $(\varphi \rightarrow \psi)$ whenever φ and ψ are in $\mathcal{L}$. Also $\mathcal{L}^{\rightarrow}$ is identified with the set of its well-formed formulas.

To give acceptability conditions for our conditional, we use Popper functions. $P : \mathcal{L} \times \mathcal{L} \rightarrow [0, 1]$ is a Popper function on $\mathcal{L} \times \mathcal{L}$ iff

1. For some $\alpha, \beta \in \mathcal{L}$, $P(\alpha|\beta) \neq 1$;

and for all $\varphi, \psi, \chi, \eta \in \mathcal{L}$,

2. If $\models_{PL} \psi \equiv \chi$, then $P(\varphi|\psi) = P(\varphi|\chi)$;

3. If $\varphi \models_{PL} \psi$, then $P(\psi|\varphi) = 1$;

4. If $\varphi \models_{PL} \neg(\psi \wedge \chi)$, then $P(\psi \vee \chi|\varphi) = P(\psi|\varphi) + P(\chi|\varphi)$ (i.e., $P(\cdot|\varphi)$ is a finitely additive probability measure) or $P(\eta|\varphi) = 1$;

5. $P(\psi \wedge \chi|\varphi) = P(\psi|\varphi)P(\chi|\psi \wedge \varphi)$.

One could define Popper functions on $\mathcal{L}$ without relying on the classical notion of logical truth/consequence (Hawthorne 1996, definition 3). We work with the above definition, however, because it makes the connection between Popper functions and unconditional probability measures clear. The latter can be recovered from Popper functions by conditionalization on $\top$. If $P(\varphi|\top) > 0$, we have

$$\frac{P(\varphi \wedge \psi|\top)}{P(\varphi|\top)} = P(\psi|\varphi \wedge \top) = P(\psi|\varphi).$$

Popper functions allow for non-trivial conditionalization on 0 probabilities: we can have that $P(\varphi|\top) = 0$ but $P(\psi|\varphi) \in (0,1)$. We call an element φ of $\mathcal{L}$ *abnormal with respect to* P when $P(\eta|\varphi) = 1$ for all $\eta \in \mathcal{L}$; *normal* otherwise.

To give the topicality component for our conditionals, we need to enrich our usual algebra of topics a bit. Let's call *topic model* $\mathfrak{T}$ a tuple $\langle \mathcal{T}, \oplus, t, f \rangle$, where $\mathcal{T}$ is our familiar set of topics, $\oplus$ is our familiar topic fusion (with topic parthood $\leq$), t assigns an item in $\mathcal{T}$ to each item in $\mathcal{L}_{AT}$ and is extended to the whole of $\mathcal{L}$ the usual way, namely with $t(\varphi) = \oplus \mathfrak{A} t \varphi$. The extra bit is $f : \mathcal{T} \to \mathcal{T}$, a function on $\mathcal{T}$ that satisfies, for all $x, y \in \mathcal{T}$:

(i) $x \leq f(x)$ [Inclusion]

(ii) $f(x) = f(f(x))$ [Idempotence]

(iii) $f(x \oplus y) = f(x) \oplus f(y)$ [Additivity]

If you recall Section 5.5, you will have spotted that f is but our Kuratowski closure operator on the partially ordered set $(\mathcal{T}, \leq)$. When it's about our indicative conditionals, we can think of it as mapping the topic of the conditional antecedent φ to the topic of the relevant background assumptions BA_φ determined by φ and context. Given this role of f, (i)-(iii) are well-motivated (compare Section 5.5 again): inclusion (i) guarantees that the topic of the relevant background assumptions BA_φ possibly expands, but always includes, the topic of the antecedent φ that triggers the conditional supposition. This constraint fits with our assumption that $\varphi \in BA_\varphi$ and allows us to account for cases such as (11)-(13). Idempotence

(ii) states that the set of background assumptions BA_φ determined by φ is complete: contemplating on the background assumptions triggered by φ does not lead to new background assumptions unless given additional inputs. Additivity (iii) ensures that the topic of the relevant background assumptions BA_φ determined by φ is the same as the fusion of the topics of the relevant background assumptions determined by its more primitive components.

This is still an abstract characterization of how antecedent φ and context conjure to determine the topic of the relevant background assumptions BA_φ. One could ask (and, a reviewer of this book did ask), exactly *what* should go in there? And one would find our account rather uninformative on this. More could be said for sure, but the situation seems to us not much worse than what happens in the mainstream Lewis-Stalnaker possible worlds semantics for *ceteris paribus* conditionals. For counterfactuals (the kind of conditionals for which it enjoys more popularity), in a simple version, this has that 'If it were the case that φ, it would be the case that ψ' is true when ψ is true throughout the closest φ-worlds, where closeness, as we know, is understood as similarity. One could ask, exactly which worlds should go in the relevant set? What are the relevant respects of similarity? The Lewis-Stalnaker account is rather uninformative on this. More could be said, and theorists subscribing to it have been talking of how to fine-tune similarity for decades; for a survey see, e.g., chs. 11-13 of Bennett (2003).

Given a topic model $\mathfrak{T} = \langle \mathcal{T}, \oplus, t, f \rangle$, a conditional $\varphi \to \psi$ is on-topic with respect to $\mathfrak{T}$ when $t(\psi) \leq f(t(\varphi))$. $\varphi \to \psi$ is on-topic *simpliciter* when it's an on-topic conditional with respect to all topic models. Being on-topic is what makes a conditional relevant: the topic of its consequent is included in that contextually determined by its antecedent, and given via f. Notice that for any topic model $\mathfrak{T}$, and $x, y \in \mathcal{T}$, if $x \leq y$ then $f(x) \leq f(y)$.[2]

We can now define the graded acceptability conditions for the formulas of $\mathcal{L}^\to$, and in particular for our indicatives. For

---

[2] *Proof*: let $\mathfrak{T} = \langle \mathcal{T}, \oplus, t, f \rangle$ be a topic model and $x, y \in \mathcal{T}$ such that $x \leq y$, that is, $x \oplus y = y$. Since f is well-defined, $f(x \oplus y) = f(y)$. Then, Additivity (iii) above guarantees that $f(x) \oplus f(y) = f(y)$, i.e., $f(x) \leq f(y)$.

any Popper function P and topic model $\mathfrak{T}$ defined on $\mathcal{L}$, the degree of acceptability $Acc_{P,\mathfrak{T}} : \mathcal{L}^{\rightarrow} \rightarrow [0,1]$ of a formula in $\mathcal{L}^{\rightarrow}$ is:

(a) For all $\varphi \in \mathcal{L}$, $Acc_{P,\mathfrak{T}}(\varphi) = P(\varphi|\top)$; and

(b) $Acc_{P,\mathfrak{T}}(\varphi \rightarrow \psi) = \begin{cases} P(\psi|\varphi), & \text{if } t(\psi) \leq f(t(\varphi)) \\ 0 & \text{otherwise.} \end{cases}$

Part (a) of our definition says that the degree of acceptability of a Boolean sentence $\varphi \in \mathcal{L}$ goes by $P(\varphi|\top)$. Notice that topic models play no role in stating the degree of acceptability of a Boolean sentence. Part (b) makes our key claim for this chapter precise: the degree of acceptability of $\varphi \rightarrow \psi$ is (i) the probability of ψ conditional on φ, as per AT, so long as (ii) $\varphi \rightarrow \psi$ is an on-topic indicative; otherwise $\varphi \rightarrow \psi$ is plainly unacceptable. This is, thus, exactly how we propose to fix AT: by adding the constraint that acceptability drops down to zero when the conditional fails to be on-topic.

One reviewer of this book asks whether this account of simple indicatives should count as semantics or as pragmatics. Well, that depends on how one wants to draw the boundary. We have rejected since Section 8.1 above the idea that relevance for indicatives boils down to traditional Gricean pragmatics: even those like Lassiter (2021), who object to making relevance part of the truth-conditional content of conditionals, grant that it's no mere cancellable conversational implicature. On the other hand, unlike, e.g., some approaches discussed in Section 8.2, we have not made of relevance (in the form of topicality-preservation) part of the truth conditional meaning of indicatives: we have not given truth conditions for them at all, only acceptability conditions. And we have done that in a way that is systematic and formalized. For some people, this may be enough to take one into semantic territory. But one may understand 'semantics' narrowly: one only has a semantics if one is sticking strictly to truth conditions. For one who thinks this way, ours is indeed not a semantics. We hope it proves to be something interesting, even if one may argue on how to categorize the something.

8.4   The Logic of On-Topic Indicatives

'Any complete theory of conditionals requires a theory of conditional inference' (Evans and Over 2004, 168). So now that we have a probabilistic and topic-sensitive account of indicatives, we set up a probabilistic logic to investigate inferences involving them.

We present the closure principles of interest as premise-conclusion rules of the form '$\Gamma \vdash \Delta$' where $\Gamma, \Delta \subseteq \mathcal{L}^{\rightarrow}$ with $\Gamma = \emptyset$ for zero-premise rules. For any Boolean $\varphi \in \mathcal{L}$, '$\vdash_{PL} \varphi$' says that φ is a theorem of classical propositional logic. Following Adams (1998), we define validity probabilistically in terms of degrees of *un*acceptability (he calls it 'uncertainty', a bit misleadingly). For any $\varphi \in \mathcal{L}^{\rightarrow}$, the degree of unacceptability $Un_{P,\mathfrak{T}}(\varphi)$ is given by $Un_{P,\mathfrak{T}}(\varphi) = 1 - Acc_{P,\mathfrak{T}}(\varphi)$.[3] When it is clear which Popper function and topic model are used, we omit the subscripts and simply write 'Acc' and 'Un' for acceptability and unacceptability respectively.

A principle of the form $\Gamma \vdash \Delta$ is valid iff for any Popper function P and topic model $\mathfrak{T}$,

$$\sum_{\varphi \in \Gamma} Un(\varphi) \geq Un(\psi)$$

for all $\psi \in \Delta$, that is, the unacceptability of the conclusion does not exceed the sum of the unacceptabilities of the premises. When $\Gamma = \emptyset$, we say $\vdash \Delta$ is valid iff $Un(\psi) = 0$ for all $\psi \in \Delta$. $\Gamma \vdash \Delta$ is invalid otherwise.

In spite of our following Adams, our logic differs from his because it embeds two factors for the (un)acceptability of conditionals: (i) probability and (ii) relevance or topicality, as per our two-component account of acceptability for indicatives. Besides investigating valid closure principles, we want to check that the invalid ones fail for the right reason. So we consider *probabilistic validity* and *topical validity* separately,

---

[3]Given a Popper function P and a topic model $\mathfrak{T}$, we have:

1. For all $\varphi \in \mathcal{L}$, $Un_{P,\mathfrak{T}}(\varphi) = 1 - P(\varphi|\top)$; and

2. $Un_{P,\mathfrak{T}}(\varphi \rightarrow \psi) = \begin{cases} 1 - P(\psi|\varphi), & \text{if } t(\psi) \leq f(t(\varphi)) \\ 1 & \text{otherwise.} \end{cases}$

and highlight the distinct sources of invalidity. However, our focus keeps being the notion of validity given in the definition just presented. We use the notions of probabilistic and topical validity in order to point out the subtle reasons for invalidity.

We say that $\Gamma \vdash \Delta$ is *probabilistically valid* (*p*-valid) iff, for any Popper function P and *singleton* topic model $\mathfrak{T}$ (i.e., topic model in which $\mathcal{T}$ is a singleton),

$$\sum_{\varphi \in \Gamma} Un(\varphi) \geq Un(\psi),$$

for all $\psi \in \Delta$. When $\Gamma = \emptyset$, we say $\vdash \Delta$ is *p-valid* iff $Un(\psi) = 0$ for all $\psi \in \Delta$; and $\Gamma \vdash \Delta$ is *p-invalid* otherwise.

We say that $\Gamma \vdash \Delta$ is *topically valid* (*t*-valid) iff, for any topic model $\mathfrak{T} = \langle \mathcal{T}, \oplus, t, f \rangle$, if every conditional in Γ is an on-topic conditional with respect to $\mathfrak{T}$ then every conditional in Δ is also an on-topic conditional wrt $\mathfrak{T}$; and $\Gamma \vdash \Delta$ is *t-invalid* otherwise.

Our *p*-validity works similarly to Adams' *p*-validity – except that we define it in terms of Popper functions instead of unconditional probability functions – and it bypasses the topicality constraint by focusing on singleton topic models (see the technical appendix of Berto and Özgün (2021) for details on how, and why, this works). *t*-validity, on the other hand, ignores probabilistic constraints and checks whether a closure principle satisfies the required relevance or topic-inclusion condition.

We now focus on the following closure principles (we label them, again, following Douven (2016), who sticks to popular names from the literature):

(REF) $\vdash \varphi \to \varphi$

(ANT) $\varphi \to \psi \vdash \varphi \to (\varphi \wedge \psi)$

(CM) $\varphi \to (\psi \wedge \chi) \vdash \varphi \to \psi, \varphi \to \chi$

(CC) $\varphi \to \psi, \varphi \to \chi \vdash \varphi \to (\psi \wedge \chi)$

(CSO) $\varphi \to \psi, \psi \to \varphi, \varphi \to \chi \vdash \psi \to \chi$

(CT) $\varphi \to \psi, (\varphi \wedge \psi) \to \chi \vdash \varphi \to \chi$

(CMon) $\varphi \to \psi, \varphi \to \chi \vdash (\varphi \wedge \psi) \to \chi$

(OR) $\varphi \to \psi, \chi \to \psi \vdash (\varphi \vee \chi) \to \psi$

(M. Ponens) $\varphi, \varphi \to \psi \vdash \psi$

(Trans) $\varphi \to \psi, \psi \to \chi \vdash \varphi \to \chi$

(SA) $\varphi \to \psi \vdash (\varphi \wedge \chi) \to \psi$

(MOD) $\neg\varphi \to \varphi \vdash \psi \to \varphi$

(RCE) If $\varphi \vdash_{PL} \psi$, then $\vdash \varphi \to \psi$

(RCEA) If $\vdash_{PL} \varphi \equiv \psi$, then $\varphi \to \chi \dashv\vdash \psi \to \chi$

(RCEC) If $\vdash_{PL} \varphi \equiv \psi$, then $\chi \to \varphi \dashv\vdash \chi \to \psi$

(RCK) If $\vdash_{PL} (\varphi_1 \wedge \cdots \wedge \varphi_n) \supset \psi$, then $\chi \to \varphi_1, \ldots, \chi \to \varphi_n \vdash \chi \to \psi$

(RCM) $\vdash_{PL} \varphi \supset \psi$, then $\chi \to \varphi \vdash \chi \to \psi$

(And-to-If) $\varphi \wedge \psi \vdash \varphi \to \psi$

(Or-to-If) $\varphi \vee \psi \vdash \neg\varphi \to \psi$

(Contr.) $\varphi \to \neg\psi \vdash \psi \to \neg\varphi$

(SDA) $(\varphi \vee \psi) \to \chi \vdash \varphi \to \chi, \psi \to \chi$

In our Berto and Özgün (2021), we proved the following (the proof is in the technical appendix to the paper):

1. REF, ANT, CM, CC, CSO, CT, CMon, OR, and Modus Ponens are both p- and t-valid. Therefore, they all are valid.

2. MOD, RCE, RCEA, RCEC, RCK, RCM, and And-to-If are p-valid but t-invalid: these are, thus, the invalidities delivered specifically by the topic-sensitivity of our conditional.

3. Trans and SA are p-invalid but t-valid.

4. Or-to-if, Contraposition, and SDA are both p-invalid and t-invalid.

5. MOD, RCE, RCEA, RCEC, RCK, RCM, And-to-If, Trans, SA, Or-to-If, Contraposition, and SDA are invalid.

We comment on some notable validities and invalidities. As for the former, REF (Reflexivity) and ANT appear fairly

	valid	p-valid	t-valid
REF	✓	✓	✓
ANT	✓	✓	✓
CM	✓	✓	✓
CC	✓	✓	✓
CSO	✓	✓	✓
CT	✓	✓	✓
CMon	✓	✓	✓
OR	✓	✓	✓
Modus Ponens	✓	✓	✓
MOD	X	✓	X
RCE	X	✓	X
RCEA	X	✓	X
RCEC	X	✓	X
RCK	X	✓	X
RCM	X	✓	X
And-to-If	X	✓	X
Trans	X	X	✓
SA	X	X	✓
Or-to-If	X	X	X
Contraposition	X	X	X
SDA	X	X	X

Table 8.1: Validities (✓) and invalidities (X): summary of the results in Berto and Özgün (2021).

obvious. CT and CMon have already been discussed above. Modus Ponens, we have argued, is desirable. The other validities hold in most conditional logics and theories of non-monotonic entailment (Nute 1984).

Let's focus on CC, and the story promised in Section 8.2 about Lottery Paradox cases. We mentioned there that one might object to the validity of CC on the basis of such cases. However, Lottery Paradox scenarios seem to rely on a more qualitative interpretation of acceptability, based on the idea that something becomes (plainly) acceptable upon passing an intermediate probabilistic threshold θ. Now our setting is not like this: it is a fully quantitative setting with degrees of acceptability. On the other hand, the most natural qualitative rephrasing of our setting does invalidate CC for $\theta \in (\frac{1}{2}, 1)$. It works thus: take a simple indicative $\varphi \rightarrow \psi$ to be (plainly) acceptable iff (i) $P(\psi|\varphi) \geq \theta$ and (ii) $t(\psi) \leq f(t(\varphi))$; define the corresponding notion of validity as Douven (2016) does: then CC becomes invalid for all threshold values $\theta \in (\frac{1}{2}, 1)$. And so we have a way of making happy those who want to stick to the invalidity of CC, at least for the relevant reading of acceptability.

The invalidities in group 2 are all related to the hyperintensional acceptability conditions of our conditional: they are p-valid; their invalidity is due to failures of topic-inclusion. Look for instance at RCEA and RCEC: that φ and ψ are classically equivalent (their material equivalence is a theorem of classical propositional logic) doesn't guarantee their replacement in the antecedent or consequent of a conditional to preserve acceptability. (RCEA and RCEC are, of course, counterparts of our Equivalence and Closure-under-$\prec$ principles for TSIMs, whose failure has been flagged since Section 3.4.) Taking '5 is prime' and 'One cannot square the circle' as necessarily equivalent (*qua* true in all possible worlds), our sample conditionals above, (9) ('If all even numbers are prime and 5 is even, then 5 is prime') and (10) ('If all even numbers are prime and 5 is even, then one cannot square the circle') are not both acceptable: only the former's consequent is on-topic with respect to the antecedent. Or, look at RCE: that φ classically entails ψ doesn't make the corresponding conditional acceptable. 'Obama is tall' entails 'Either it is raining in Melbourne or not' in classical logic, but we don't accept 'If Obama is tall, then either it is raining in Melbourne or not', as the latter is patently off-topic. We find here, in the probabilistic-conditional setting, the same hyperintensional patterns we found in (the non-probabilistic semantics for) our two-place TSIMs in previous chapters.

Also (the validity of) CSO from group 1 should look familiar: it's the counterpart of our Restricted Equivalence principle for two-place TSIMs, discussed in chapters 5 and 6. Just as Restricted Equivalence limited the hyperintensional anarchy of those TSIMs, so does CSO limit the hyperintensional anarchy of indicatives: even if replacement of necessary or logical equivalents fails to preserve acceptability, CSO guarantees that replacement of 'conditional equivalents' does: when both $\varphi \to \psi$ and the converse $\psi \to \varphi$ are among the premises, the inference from these and $\varphi \to \chi$ to the conditional obtained by replacing φ with ψ in the latter, namely $\psi \to \chi$, is valid.

Groups 3 and 4 include inferences generally agreed to be invalid for any *ceteris paribus* conditional in the indicative and even in the subjunctive-counterfactual mood: SA (Strengthening the Antecedent), Contraposition, Transitivity, SDA (Simplification of Disjunctive Antecedents), Or-to-If, fail both in the Adams (1998) probabilistic semantics for indicatives and in the possible worlds semantics for indicatives and/or counterfactuals by Stalnaker (1968) and Lewis (1973).

Finally, And-to-If fails in the most natural way. Notice the precise sense in which it does. We don't claim that the inference fails to be truth-preserving (we have not given truth conditions for indicatives at all). Nor is the inference probabilistically invalid in the sense of p-validity: it is, instead, p-valid, just as in the Adams setting. It is not t-valid, however. The inference from $\varphi \wedge \psi$ to $\varphi \to \psi$ fails to be acceptability-preserving due to the topicality constraint: the latter may be an off-topic conditional like our (5) above ('If raccoons have no wings, then they cannot breath under water') although the former is a true and acceptable conjunction like our (4) ('Raccoons have no wings and they cannot breath under water'). Although the conjuncts plausibly overlap in topic (they are both about raccoons), which makes the conjunction acceptable, and coherently assertable in natural conversational contexts, the topic of 'Raccoons cannot breath under water' is not fully included in that of the background assumptions contextually triggered by 'Raccoons have no wings'.

Such (in)validities make for a conditional logic that is not only theoretically desirable, but also *empirically* plausible. As noted, e.g., in Evans and Over (2004), 44-5, the vast majority of experimental results concerning how people reason with conditionals only involve four simple inferences: Modus Ponens, Modus Tollens, and the usual fallacies of Affirming the Consequent and Denying the Antecedent. There are few studies investigating other conditional inferences (we mentioned Pfeifer and Kleiter (2010) above as one notable exception). However, an initial and tentative assessment of the psychological plausibility of our logic is possible, thanks to a sophisticated experiment reported in Douven (2016), ch. 5.

Acknowledging that many inferences considered in the literature on conditional logics, including various among those in our table, are of a kind that people would rarely make in everyday reasoning, Douven went on to test them experimentally in a more roundabout way. Here's a summary of what he did (the detailed presentation is on pp. 140ff. of Douven's book).

Most inferences involving simple conditionals in conditonal logic feature at most three propositions (expressed by sentences) φ, ψ, and χ. So Douven asked over 1,000 subjects to rate the probabilities of conjunctions of the form $\pm\varphi\wedge\pm\psi\wedge\pm\chi$ (called *atoms*), with '$\pm$\$' indicating that sentence \$ may occur negated or unnegated, and φ, ψ, χ taken from news websites. For each triple of sentences, there are eight mutually exclusive and jointly exhaustive combinations (exactly one of them has to be true), composing an *atom matrix*. Subjects were instructed that the truth of any element of the matrix would exclude that of all the others, and that one of the elements had to be true, so that the probabilities assigned to the atoms had to add up to 100%.

Douven then computed which conditional-involving inferences with φ, ψ, χ end up acceptability-preserving. He checked acceptability-preservation for two thresholds, $\theta = 0.5$ and $\theta = 0.9$. The results summarized in Douven (2016), 144, show that the validities of our logic tested in the experiment correspond to highly popular inferences: CC has percentages of 100/100% endorsement (for 0.5 and 0.9 respectively); CSO has 75/100%; CT has 87/94%; CMon has 86/100%; Modus Ponens has 91/78%. Vice versa, some invalidities have low endorsement rates: Or-to-If has 28/9%; SDA has 44/56%.

An open issue is that the three inferences of Contraposition (70/78%), SA (76/97%) and Transitivity (78/100%) are highly endorsed. However, these are invalid, as we mentioned, in virtually any conditional logic for non-monotonic and *ceteris paribus* conditionals. Their invalidity, furthermore, is not due specifically to the distinctive element of our semantics for the indicative, namely our topicality constraint: they are invalidated purely probabilistically in semantics à la Adams, and they fail also in similarity-based possible worlds semantics à la Stalnaker-Lewis, due to conditionals being variably strict in the approach.

Unsurprisingly, therefore, authors endorsing some variant of any of these treatments of conditionals have come up with explanations for the popuarity of such invalid inferences. In particular, the fact that the three of them are more popular with higher 0.9 threshold than with lower 0.5, may corroborate the story proposed in Adams (1998); Bennett (2003): such inferences fail for non-perfectly-certain propositions, so it is plausible that their endorsement grows as we lift the threshold towards certainty, i.e., probability 1. They tend to be endorsed to the extent that they are mistaken for their (themselves popular) limited counterparts, like Cautious Monotonicity (CMon) and Cautious Transitivity (CT).

We have focused on closure principles that are more commonly discussed in the context of conditional logics and non-monotonic reasoning, and empirically tested by Douven (2016). A more exhaustive list can be found, e.g., in Douven (2016), 129, and Crupi and Iacona (2021), 6,. How to extend our analysis for those additional principles should be obvious. We can adopt the components of Crupi and Iacona (2021)'s framework concerning the operators necessity ($\Box$), possibility ($\Diamond$), and negation ($\sim$), and evaluate the principles involving them with respect to our topic-sensitive semantics.

8.5   Chapter Summary

This chapter has combined topic-sensitivity and probabilities to provide an account of (simple) indicative conditionals. The account is in the spirit of Adams' Thesis, in that the acceptability of a simple indicative is tied to the corresponding conditional probability. But it fixes the empirical and theoretical shortcomings of the Thesis by adding a relevance constraint for acceptability, where relevance is understood, again, as topic-sensitivity. The chapter has presented a probabilistic logic for simple indicatives in terms of Popper functions, arguing that its (in)validities are both plausible and in line with empirical results on how people reason with conditionals.

Bibliography

Adams, E. (1966). Probability and the logic of conditionals, in J. Hintikka and P. Suppes (eds), *Aspects of Inductive Logic*, North-Holland, Amsterdam, pp. 165–316.

Adams, E. (1975). *The Logic of Conditionals*, Reidel, Dordrecht.

Adams, E. (1998). *A Primer of Probability Logic*, CSLI Publications, Stanford, CA.

Alchourrón, C., Gärdenfors, P. and Makinson, D. (1985) On the logic of theory change: Partial meet functions for contraction and revision, *Journal of Symbolic Logic* 50: 510–30.

Anderson, A. and Belnap, N. (1975). *Entailment: The Logic of Relevance and Necessity*, Vol. *I*, Princeton University Press, Princeton, NJ.

Anderson, J. (1983). *The Architecture of Cognition*, Harvard University Press, Cambridge, MA.

Angell, R. (1977). Three systems of first degree entailment, *Journal of Symbolic Logic* 47: 147.

Arcangeli, M. (2019). *Supposition and the Imaginative Realm*, Routledge, New York.

Arlo-Costa, H. and Parikh, R. (2005). Conditional probability and defeasible inference, *Journal of Philosophical Logic* 34: 97–119.

Armstrong, D. M. (2004). *Truth and Truthmakers*, Cambridge University Press, Cambridge.

Asheim, G. and Sovik, Y. (2005). Preference-based belief operators, *Mathematical Social Sciences* 50: 61–82.

Atance, C. and O'Neill, D. (2001). Episodic future thinking, *Trends in Cognitive Science* 5: 533–39.

Avnur, Y., Brueckner, A. and Buford, C. (2011). No closure on skepticism, *Pacific Philosophical Quarterly* 92: 439–47.

Baddeley, A. (1986). *Working Memory*, Oxford University Press, New York.

Baddeley, A. (2002). Is working memory still working?, *European Psychologist* 7: 85–97.

Badura, C. (2021a). How imagination can justify, in C. Badura and E. Kind (eds), *Epistemic Uses of Imagination*, Routledge, New York.

Badura, C. (2021b). More aboutness in imagination, *Journal of Philosophical Logic* 50: 523–47.

Badura, C. and Wansing, H. (2021). Stit-logic for imagination episodes with voluntary input, *Review of Symbolic Logic* doi:10.1017/S1755020321000514.

Balbiani, P., Fernández-Duque, D. and Lorini, E. (2019). The dynamics of epistemic attitudes in resource-bounded agents, *Studia Logica* 107: 457–88.

Balcerak Jackson, M. (2016). On the epistemic value of imagining, supposing and conceiving, in A. Kind and P. Kung (eds), *Knowledge through Imagination*, Oxford University Press, Oxford, pp. 42–60.

Baltag, A. Moss, L. and Solecki, S. (1998). The logic of public announcements, common knowledge, and private suspicions, in I. Gilboa (ed.), *Proceedings of TARK 98*, Morgan and Kaufmann, Evanston, IL, pp. 43–56.

Baltag, A. and Renne, B. (2016). Dynamic epistemic logic, in E. N. Zalta (ed.), *The Stanford Encyclopedia of Philosophy*, winter 2016 edn, Metaphysics Research Lab, Stanford University, Stanford, CA.

Baltag, A. and Smets, S. (2008a). Probabilistic dynamic belief revision, *Synthese* 165: 143–66.

Baltag, A. and Smets, S. (2008b). A qualitative theory of dynamic interactive belief revision, in G. Bonanno, W. van der Hoek and M. Wooldridge (eds), *Logic and the Foundations of Game and Decision Theory*, Amsterdam University Press, Amsterdam, pp. 9–58.

Barsalou, L. (1992). *Cognitive Psychology*, Erlbaum, Hillsdale, NJ.

Barwise, J. and Etchemendy, J. (1987). *The Liar: An Essay on Truth and Circularity*, Oxford University Press, Oxford.

Barwise, J. and Perry, J. (1983). *Situations and Attitudes*, CSLI Publications, Stanford, CA.

Barwise, J. and Seligman, J. (1995). *Information Flow, The Logic of Distributed Systems*, Cambridge University Press, Cambridge.

Beall, J. (2015). Free of detachment: Logic, rationality, and gluts, *Noûs* 49: 410–23.

Beall, J. C. (2016). Off-topic: A new interpretation of weak kleene logic, *Australasian Journal of Logic* 13: 136–42.

Bennett, J. (2003). *A Philosophical Guide to Conditionals*, Oxford University Press, Oxford.

Berto, F. (2012). *Existence as a Real Property*, Synthese Library, Springer, Dordrecht.

Berto, F. (2014). On conceiving the inconsistent, *Proceedings of the Aristotelian Society* 114: 103–21.

Berto, F. (2017a). Aboutness in imagination, *Philosophical Studies* 175: 1871–86.

Berto, F. (2017b).   Impossible worlds and the logic of imagination, *Erkenntnis* 82: 1277–97.

Berto, F. (2018).   Simple hyperintensional belief revision, *Erkenntnis* 84: 559–75.

Berto, F. (2021).   Equivalence in imagination, in C. Badura and E. Kind (eds), *Epistemic Uses of Imagination*, Routledge, New York.

Berto, F. and Jago, M. (2019).   *Impossible Worlds*, Oxford University Press, Oxford.

Berto, F. and Nolan, D. (2021).   Hyperintensionality, in E. N. Zalta (ed.), *The Stanford Encyclopedia of Philosophy*, spring 2021 edn, Metaphysics Research Lab, Stanford University, Stanford, CA.

Berto, F. and Özgün, A. (2021). The logic of framing effects. Unpublished Manuscript.

Berto, F. and Schoonen, T. (2018).   Conceivability and possibility: Some dilemmas for humeans, *Synthese* 195: 2697–715.

Block, N. (1983). Mental pictures and cognitive science, *The Philosophical Review* 92: 499–541.

Board, O. (2004). Dynamic interactive epistemology, *Games and Economic Behaviour* 49: 49–80.

Boghossian, P. (1996).   Analyticity reconsidered, *Noûs* 30: 360–91.

Braine, M. (1978). On the relation between the natural logic of reasoning and standard logic, *Psychological Review* 85: 1–21.

Braine, M. and O'Brien, D. (1991). A theory of *If*: Lexical entry, reasoning program and pragmatic principles, *Psychological Review* 98: 182–203.

Branquinho, J. (1990).   Are Salmon's 'guises' disguised Fregean senses?, *Analysis* 50: 19–24.

Brown, J. (2018). *Fallibilism, Evidence and Knowledge*, Oxford University Press, Oxford.

Burgess, J. (2009). *Philosophical Logic*, Princeton University Press, Princeton, NJ.

Busby, E., Flynn, D. and Druckman, J. (2018). Studying framing effects on political preferences, in P. D' Angelo (ed.), *Doing News Framing Analysis*, Vol. II, Routledge, New York, pp. 67–90.

Byrne, A. (2007). Possibility and imagination, *Philosophical Perspectives* 21: 125–44.

Byrne, R. (2005). *The Rational Imagination. How People Create Alternatives to Reality*, MIT Press, Cambridge, MA.

Byrne, R. and Girotto, V. (2009). Cognitive processes in counterfactual thinking, in K. Markman, W. Klein and J. Suhr (eds), *Handbook of Imagination and Mental Simulation*, Taylor and Francis, New York, pp. 151–60.

Canavotto, I., Berto, F. and A., G. (2020). Voluntary imagination: A fine-grained analysis, *Review of Symbolic Logic* Online First: doi 10.1017/S1755020320000039.

Chalmers, D. (2002). Does conceivability entail possibility?, in T. Gendler and J. Hawthorne (eds), *Conceivability and Possibility*, Oxford University Press, Oxford, pp. 145–99.

Chalmers, D. (2011). Propositions and attitude ascriptions: a Fregean account, *Noûs* 45: 595–639.

Chellas, B. (1975). Basic conditional logic, *Journal of Philosophical Logic* 4: 133–53.

Chellas, B. F. (1980). *Modal Logic*, Cambridge University Press, Cambridge.

Cherniak, C. (1986). *Minimal Rationality*, Bradford Books, MIT Press, Cambridge, MA.

Chierchia, G. and McConnell-Ginet, S. (1990). *Meaning and Grammar. An Introduction to Semantics*, MIT Press, Cambridge, MA.

Ciuni, R., Szmuc, D. and Ferguson, T. (2018). Relevant logics obeying component homogeneity, *Australasian Journal of Logic* 17: 301–61.

Cohen, S. (1988). How to be a fallibilist, *Philosophical Perspectives* 2: 91–123.

Cohen, S. (2002). Basic knowledge and the problem of easy knowledge, *Philosophy and Phenomenological Research* 65: 309–29.

Correia, F. (2004). Semantics for analytic containment, *Studia Logica* 77: 87–104.

Crane, T. (2013). *The Objects of Thought*, Oxford University Press, Oxford.

Crimmins, M. (1992). *Talk About Beliefs*, MIT Press, Cambridge, MA.

Crowder, R. (1993). Short-term memory: Where do we stand?, *Memory and Cognition* 21: 142–5.

Crupi, V. and Iacona, A. (2021). Three ways of being non-material, *Studia Logica* Online First: doi 10.1007/s11225-021-09949-y.

Cruse, D. (2017). *Meaning in Language*, Oxford University Press, Oxford.

Currie, G. (1990). *The Nature of Fiction*, Cambridge University Press, Cambridge.

Currie, G. and Ravenscroft, I. (2002). *Recreative Minds*, Oxford University Press, Oxford.

Dabrowski, A., Moss, L. and Parikh, R. (1996). Topological reasoning and the logic of knowledge, *Annals of Pure and Applied Logic* 78: 73–110.

Dancygier, B. and Sweetser, E. (2005). *Mental Spaces in Grammar: Conditional Constructions*, Cambridge University Press, Cambridge.

Davies, J. (2019). *Imagination. The Science of Your Mind's Greatest Power*, Pegasus Books, New York.

Declerck, R. and Reed, S. (2001). *Conditionals: A Comprehensive Empirical Analysis*, De Gruyter, Berlin.

Deutsch, H. (1979). The completeness of S, *Studia Logica* 38: 137–47.

Deutsch, H. (1984). Parconsistent analytic implication, *Journal of Philosophical Logic* 13: 1–11.

Divers, J. (2002). *Possible Worlds*, Routledge, New York.

Dorsch, F. (2012). *The Unity of Imagining*, Ontos Verlag, Frankfurt.

Douven, I. (2016). *The Epistemology of Indicative Conditionals: Formal and Empirical Approaches*, Cambridge University Press, Cambridge.

Douven, I. and Verbrugge, S. (2010). The Adams family, *Cognition* 117. 302 18.

Dretske, F. (1970). Epistemic operators, *Journal of Philosophy* 67: 1007–23.

Dretske, F. (1999). *Knowledge and the Flow of Information*, CLSI Publications, Stanford, CA.

Dretske, F. (2005). The case against closure, in M. Steup and E. Sosa (eds), *Contemporary Debates in Epistemology*, Blackwell, Oxford, pp. 13–26.

Druckman, J. (2001a). Evaluating framing effects, *Journal of Economic Psychology* 22: 96–101.

Druckman, J. (2001b). Using credible advice to overcome framing effects, *Journal of Law, Economics, and Organization* 17: 62–68.

Dunn, J. (1972). A modification of parry's analytic implication, *Notre Dame Journal of Formal Logic* 13: 195–205.

Dunn, J. and Restall, G. (2002). Relevance logic, in D. Gabbay and F. Guenthner (eds), *Handbook of Philosophical Logic*, second edn, Vol. 6, Kluwer Academic, Dordrecht, pp. 1–136.

Edgington, D. (1995). On conditionals, *Mind* 104: 235–329.

Elqayam, S. and Over, D. (2013). New paradigm psychology of reasoning, *Thinking and Reasoning* 19: 249–65.

Epstein, R. (1981). Relatedness and dependence in propositional logics, *Journal of Symbolic Logic* 46: 202–3.

Epstein, R. (1993). *The Semantic Foundations of Logic: Propositional Logics*, Oxford University Press, New York.

Evans, J., Handley, S. and Over, D. (2003). Conditionals and conditional probability, *Journal of experimental psychology. Learning, memory, and cognition* 29: 321–35.

Evans, J. and Over, D. (2004). *If*, Oxford University Press, Oxford.

Eysenck, M. and Keane, M. (2015). *Cognitive Psychology*, Psychology Press, New York.

Fagin, R. and Halpern, J. (1988). Belief, awareness and limited reasoning, *Artificial Intelligence* 34: 39–76.

Fagin, R., Halpern, J., Moses, Y. and Vardi, M. (1995). *Reasoning about Knowledge*, MIT Press, Cambridge, MA.

Felka, K. (2018). Comments on Stephen Yablo's *Aboutness*, *Erkenntnis* 83: 1181–94.

Ferguson, T. (2014). A computational interpretation of conceptivism, *Journal of Applied Non-Classical Logics* 24: 333–67.

Field, H. (1978). Mental representation, *Erkenntnis* 13: 9–61.

Field, H. (1986a). Critical notice: Robert Stalnaker, *Inquiry*, *Philosophy of Science* 53: 425–48.

Field, H. (1986b). Stalnaker on intentionality: On Robert Stalnaker's *Inquiry, Pacific Philosophical Quarterly* 67: 98–112.

Fine, K. (1986). Analytic implication, *Notre Dame Journal of Formal Logic* 27: 169–79.

Fine, K. (2016a). Angellic content, *Journal of Philosophical Logic* 45: 199–226.

Fine, K. (2016b). Constructing the impossible, in L. Walters and J. Hawthorne (eds), *Conditionals, Probability, and Paradox: Themes from the Philosohy of Dorothy Edgington*, Oxford University Press, Oxford.

Fine, K. (2017). A theory of truthmaker content I: Conjunction, disjunction and negation, *Journal of Philosophical Logic* 46: 625–74.

Fine, K. (2020). Yablo on subject matter, *Philosophical Studies* 177: 129–71.

Fine, K. and Jago, M. (2018). Logic for exact entailment, *Review of Symbolic Logic* 12: 536–56.

Fiocco, M. (2007). Conceivability, imagination and modal knowledge, *Philosophy and Phenomenological Research* 74: 364–80.

Floridi, L. (2019). Semantic conceptions of information, in E. N. Zalta (ed.), *The Stanford Encyclopedia of Philosophy*, Winter 2019 edn, Metaphysics Research Lab, Stanford University, Stanford, CA.

Foley, R. (1993). *Working Without a Net*, Oxford University Press, Oxford.

Forbes, G. (1987). Review of *Freges Puzzle*, by Nathan Salmon, *Philosophical Review* 96: 455–8.

Gabbay, D. (1985). *Theoretical Foundations for Non-monotonic Reasoning in Expert Systems*, Springer, Berlin.

Gächter, S., Orzen, H., Renner, E. and Stamer, C. (2009). Are experimental economists prone to framing effects? a natural field experiment, *Journal of Economic Behavior and Organization* 70: 443–6.

Gemes, K. (1994). A new theory of content I: Basic content, *Journal of Philosophical Logic* 23: 595–620.

Gemes, K. (1997). A new theory of content II: Model theory and some alternatives, *Journal of Philosophical Logic* 26: 449–76.

Gendler, T. (2000a). The puzzle of imaginative resistance, *Journal of Philosophy* 97: 55–81.

Gendler, T. (2000b). *Thought Experiments: On the Powers and Limits of Imaginary Cases.*, Routledge, London-New York.

Gendler, T. (2020). Imaginative resistance, in E. N. Zalta (ed.), *The Stanford Encyclopedia of Philosophy*, summer 2020 edn, Metaphysics Research Lab, Stanford University, Stanford, CA.

Gillies, A. (2004). Epistemic conditionals and conditional epistemics, *Noûs* 38: 585–616.

Giordani, A. (2019). Axiomatizing the logic of imagination, *Studia Logica* 107: 639–57.

Girard, P. and Rott, H. (2014). Belief revision and dynamic logic, in A. Baltag and S. Smets (eds), *Johan Van Benthem on Logic and Information Dynamics*, Springer, Dordrecht, pp. 203–33.

Girotto, V. and Johnson-Laird, P. (2010). Conditionals and probability, in M. Oaksford and N. Chater (eds), *Cognition and Conditionals*, Oxford University Press, Oxford, pp. 103–15.

Goldman, A. (2002). *Simulating Minds*, Oxford University Press, Oxford.

Goodman, N. (1961). About, *Mind* 70: 1–24.

Gregory, D. (2016). Imagination and mental imagery, in A. Kind (ed.), *The Routledge Handbook of Philosophy of Imagination*, Routledge, New York, pp. 97–110.

Grice, H. (1989). *Studies in the Way of Words*, Harvard University Press, Cambridge, MA.

Grove, A. (1988). Two modellings for theory change, *Journal of Philosophical Logic* 17: 157–70.

Haegeman, L. (2003). Conditional clauses: External and internal syntax, *Mind and Language* 18: 317–39.

Hajek, D. (1989). Probabilities of conditionals, *Journal of Philosophical Logic* 18: 423–8.

Halpern, J. (2001). Alternative semantics for unawareness, *Games and Economic Behavior* 37: 321–39.

Halpern, J. (2005). *Reasoning About Uncertainty*, MIT Press, Cambridge, MA.

Hansson, S. (1999). *A Textbook of Belief Dynamics. Theory Change and Database Updating*, Kluwer, Dordrecht.

Harman, G. (1973). *Thought*, Princeton University Press, Princeton, NJ.

Harman, G. (1986). *Change In View*, MIT Press, Cambridge, MA.

Harman, G. and Sherman, B. (2004). Knowledge, assumption, lotteries, *Philosophical Issues* 14: 492–500.

Hawke, P. (2016). Questions, topics and restricted closure, *Philosophical Studies* 73: 2759–84.

Hawke, P. (2017). *The problem of epistemic relevance*, PhD thesis, Stanford University.

Hawke, P. (2018). Theories of aboutness, *Australasian Journal of Philosophy* 96: 697–723.

Hawke, P., Özgün, A. and Berto, F. (2020). The fundamental problem of logical omniscience, *Journal of Philosophical Logic* 49: 727–66.

Hawthorne, J. (1996). On the logic of nonmonotonic conditionals and conditional probabilities, *Journal of Philosophical Logic* 25: 185–218.

Hawthorne, J. (2004). *Knowledge and Lotteries*, Oxford University Press, Oxford.

Hawthorne, J. and Yli-Vakkuri, J. (2020). Being in a position to know, *Philosophical Studies* doi.org/10.1007/s11098-021-01709-x.

Heim, I. and Kratzer, A. (1997). *Semantics in Generative Grammar*, Wiley-Blackwell, Oxford.

Hintikka, J. (1962). *Knowledge and Belief. An Introduction to the Logic of the Two Notions*, Cornell University Press, Ithaca, NY.

Hoek, D. (2022). Minimal rationality and the web of questions, in D. Kindermann, P. Van Elswyk and E. Egan (eds), *Unstructured Content*, Oxford University Press, Oxford.

Holliday, W. (2012). *Knowing what follows: epistemic closure and epistemic logic*, PhD Dissertation, Stanford, CA.

Holliday, W. (2015). Fallibilism and multiple paths to knowledge, in T. Gendler and J. Hawthorne (eds), *Oxford Studies in Epistemology*, Vol. 5, Oxford University Press, Oxford, pp. 97–144.

Hornischer, L. (2017). *Hyperintensionality and synonymy*, MSc Master of Logic Dissertation, University of Amsterdam, Amsterdam.

Humberstone, L. (2008). Parts and partitions, *Theoria* 66: 41–82.

Jackson, F. (1979). On assertion and indicative conditionals, *The Philosophical Review* 88: 565–89.

Jackson, F. (1987). *Conditionals*, Blackwell, Oxford.

Jago, M. (2014). *The Impossible. An Essay on Hyperintensionality*, Oxford University Press, Oxford.

Jago, M. (2018). *What Truth Is*, Oxford University Press, Oxford.

Johnson-Laird, P. and Byrne, R. (1991). *Deduction*, Erlbaum, Mahwah, NJ.

Johnson-Laird, P. and Byrne, R. (2002). Conditionals: a theory of meaning, pragmatics and inference, *Psychological Review* 109: 646–78.

Johnson, M. and Raye, C. (1981). Reality monitoring, *Psychological Review* 88: 67–85.

Joyce, J. (1999). *The Foundations of Causal Decision Theory*, Cambridge University Press, Cambridge.

Kahneman, D. (2011). *Thinking: Fast and Slow*, Penguin, London.

Kahneman, D. and Tversky, A. (1984). Choices, values, and frames, *American Psychologist* 39: 341–50.

Khoo, J. (2016). Probabilities of conditionals in context, *Linguistics and Philosophy* 39: 1–43.

Kind, A. (2001). Putting the image back in imagination, *Philosophy and Phenomenological Research* 62: 85–109.

Kind, A. (2016). Imagining under constraints, in A. Kind and P. Kung (eds), *Knowledge through Imagination*, Oxford University Press, Oxford, pp. 145–59.

Kind, A. and Kung, P. (eds) (2016). *Knowledge through Imagination*, Oxford University Press, Oxford.

King, J. (1996). Structured propositions and sentence structure, *Journal of Philosophical Logic* 25: 495–521.

Konolige, K. (1986). What awareness isn't: a sentential view of implicit and explicit belief, in J. Halpern (ed.), *Theoretical Aspects of Reasoning About Knowledge*, Morgan Kaufmann, San Francisco, pp. 241–50.

Kosslyn, S. and Moulton, S. (2009). Mental imagery and implicit memory, in K. Markman, W. Klein and J. Suhr (eds), *Handbook of Imagination and Mental Simulation*, Taylor and Francis, New York, pp. 35–52.

Kraus, S., Lehmann, D. and Magidor, M. (1990). Nonmonotonic reasoning, preferential models and cumulative logics, *Artificial Intelligence* 44: 167–207.

Kripke, S. (1980). *Naming and Necessity*, Harvard University Press, Cambridge, MA.

Kripke, S. (2011a). Nozick on knowledge, *Philosophical Troubles: Collected Papers, Vol. 1*, Oxford University Press, Oxford.

Kripke, S. (2011b). On two paradoxes of knowledge, *Philosophical Troubles: Collected Papers, Vol. 1*, Oxford University Press, Oxford.

Krzyżanowska, K. (2015). *Between If and Then*, PhD thesis, University of Groningen.

Krzyżanowska, K., Collins, P. and Hahn, U. (2017). Between a conditional's antecedent and its consequent: Discourse coherence vs. probabilistic relevance, *Cognition* 164: 199–205.

Kung, P. (2014). You really do imagine it: Against error theories of imagination, *Noûs* 50: 90–120.

Kvanvig, J. (2006). *The Knowability Paradox*, Oxford University Press, Oxford.

Kyburg, H. (1961). *Probability and the Logic of Rational Belief*, Wesleyan University Press, Middletown CT.

Langland-Hassan, P. (2016). On choosing what to imagine, in A. Kind and P. Kung (eds), *Knowledge through Imagination*, Oxford University Press, Oxford, pp. 61–84.

Lasonen-Aarnio, M. (2014). The dogmatism puzzle, *Australasian Journal of Philosophy* 92: 417–32.

Lassiter, D. (2021). Decomposing relevance in conditionals, *Mind and Language* Forthcoming.

Lawlor, K. (2005). Living without closure, *Grazer Philosophische Studien* 697: 25–49.

Leitgeb, H. (2012). A probabilistic semantics for counterfactuals - Part A, *The Review of Symbolic Logic* 5: 26–84.

Leitgeb, H. (2017). *The Stability of Belief. How Rational Belief Coheres with Probability*, Oxford University Press, Oxford.

Leitgeb, H. and Segerberg, K. (2005). Dynamic doxastic logic: Why, how, and where to?, *Synthese* 155: 167–90.

Levesque, H. (1984). A logic of implicit and explicit belief, *National Conference on AI* AAAI-84: 198–202.

Levin, I., Gaeth, G., Schreiber, J. and Lauriola, M. (2002). A new look at framing effects: Distribution of effect sizes, individual differences, and independence of types of effects, *Organizational Behavior and Human Decision Processes* 88: 411–29.

Lewis, D. (1973). *Counterfactuals*, Blackwell, Oxford.

Lewis, D. (1976). Probabilities of conditionals and conditional probabilities, *Philosophical Review* 95: 581–9.

Lewis, D. (1988a). Relevant implication, *Theoria* 54: 161–74.

Lewis, D. (1988b). Statements partly about observation, *Philosophical Papers* 17: 1–31.

Lindström, S. and Rabinowicz, W. (1999). DDL unlimited: Dynamic doxastic logic for introspective agents, *Erkenntnis* 50: 353–85.

Lycan, W. (1989). Logical constants and the glory of truth-conditional semantics, *Notre Dame Journal of Formal Logic* 30: 390–401.

Lyons, J. (1996). *Linguistic Semantics*, Cambridge University Press, Cambridge.

MacFarlane, J. (2017). Logical constants, in E. N. Zalta (ed.), *The Stanford Encyclopedia of Philosophy*, winter 2017 edn, Metaphysics Research Lab, Stanford University, Stanford, CA.

Makinson, D. (1965). The paradox of the preface, *Analysis* 25: 205–07.

Mares, E. (2004). *Relevant Logic: A Philosophical Interpretation*, Cambridge University Press, Cambridge.

Mares, E. and Fuhrmann, A. (1995). A relevant theory of conditionals, *Journal of Philosophical Logic* 24: 645–65.

Markman, K., Klein, W. and Surh, J. (eds) (2009). *Handbook of Imagination and Mental Simulation*, Taylor and Francis, New York.

Marshall, D. (2021). Causation and fact granularity, *Synthese* doi 10.1007/s11229-021-03151-2.

McGee, V. (1985). A counterexample to modus ponens, *Journal of Philosophy* 82: 462–71.

McGee, V. (1986). Conditional probabilities and compounds of conditionals, *The Philosophical Review* 98: 485–541.

Miyake, A. and Shah, P. (1999). *Models of Working Memory*, Cambridge University Press, Cambridge.

Moltmann, F. (2018). An object-based truthmaker semantics for modals, *Philosophical Issues* 28: 255–88.

Moss, L. and Parikh, R. (1992). Topological reasoning and the logic of knowledge, *Proceedings of the 4th TARK*, Morgan Kaufmann, Evanston, IL, pp. 95–105.

Moss, S. (2018). *Probabilistic Knowledge*, Oxford University Press, Oxford.

Moss, S. (2019). Full belief and loose speech, *Philosophical Review* 128: 255–91.

Mulligan, K. (1999). La varietà e l'unità dell'immaginazione, *Rivista di estetica* 11: 53–67.

Murphy, M. (2003). *Semantic Relations and the Lexicon,* Cambridge University Press, Cambridge.

Nelson, M. (2019). Propositional attitude reports, in E. N. Zalta (ed.), *The Stanford Encyclopedia of Philosophy,* spring 2019 edn, Metaphysics Research Lab, Stanford University, Stanford, CA.

Nichols, S. (2006). *The Architecture of Imagination: New Essays on Pretence, Possibility, and Fiction,* Oxford University Press, Oxford.

Nichols, S. and Stich, S. (2003). *Mindreading. An Integrated Account of Pretence, Self-Awareness, and Understanding Other Minds,* Oxford University Press, Oxford.

Nozick, R. (1981). *Philosophical Explanations,* Harvard University Press, Cambridge, MA.

Nute, D. (1984). Conditional logic, in D. Gabbay and F. Guenthner (eds), *Handbook of Philosophical Logic: Vol. II: Extensions of Classical Logic,* Springer, Dordrecht.

Oaksford, M. (2005). Reasoning, in N. Braisby and M. Gellatly (eds), *Cognitive Psychology,* Oxford University Press, Oxford, pp. 366–92.

Oaksford, M. and Chater, N. (2010). *Cognition and Conditionals. Probability and Logic in Human Thinking,* Oxford University Press, Oxford.

Over, D. (2009). New paradigm psychology of reasoning, *Thinking and Reasoning* 15: 431–38.

Özgün, A. and Berto, F. (2020). Dynamic hyperintensional belief revision, *The Review of Symbolic Logic* Online First: doi 10.1017/S1755020319000686.

Özgün, A. and Cotnoir, A. (2021). Imagination and mereotopology. Unpublished Manuscript.

Pacuit, E. (2017). *Neighbourhood Semantics for Modal Logic,* Springer, Dordrecht.

Paivio, A. (1986). *Mental Representation*, Oxford University Press, Oxford.

Parry, W. (1933). Ein axiomensystem fr eine neue art von implikation (analitische implikation), *Ergebnisse eines Mathematischen Kolloquiums* 4: 5–6.

Parry, W. T. (1968). The logic of C.I. Lewis, in P. Schlipp (ed.), *The Philosophy of C.I. Lewis*, Cambridge University Press, Cambridge, pp. 115–54.

Parry, W. T. (1989). Analytic implication: its history, justification and varieties, in Norman and Sylvan (eds), *Directions in Relevant Logic*, Kluwer Academic Publishers, Dordrecht, pp. 101–18.

Perry, J. (1989). Possible worlds and subject matter, *The Problem of the Essential Indexical and Other Essays*, CSLI publications, Stanford, CA, pp. 145–60.

Pfeifer, N. and Kleiter, G. (2010). Conditionals and probability, in M. Oaksford and N. Chater (eds), *The Conditional in Mental Probability Logic*, Oxford University Press, Oxford, pp. 153–173.

Pinker, S. (1980). Mental imagery and the third dimension, *Journal of Experimental Psychology* 109: 354–71.

Plebani, M. (2020). Why aboutness matters: Meta-fictionalism as a case study, *Philosophia* Online First: doi 10.1007/s11406-020-00272-9.

Plebani, M. and Spolaore, G. (2021). Subject matter: A modest proposal, *The Philosophical Quarterly* 71: 605–22.

Plous, S. (1993). *The Psychology of Judgment and Decision Making*, McGraw-Hill, New York.

Pollock, J. (1994). Justification and defeat, *Artificial Intelligence* 67: 377–407.

Priest, G. (2008). *An Introduction to Non-Classical Logic. From If to Is*, Cambridge University Press, Cambridge.

Priest, G. (2016). *Towards Non-Being: The Logic and Metaphysics of Intentionality, 2nd ed.*, Oxford University Press, Oxford.

Putnam, H. (1958). Formalization of the concept 'about', *Philosophy of Science* 25: 15–130.

Pylyshyn, Z. (1981). The imagery debate: Analogue media versus tacit knowledge, *Psychological Review* 88: 16–45.

Pylyshyn, Z. (2002). Mental imagery: In search of a theory, *Behavioral and Brain Sciences* 25: 157–82.

Quine, W. (1960). *Word and Object*, MIT Press, Cambridge, MA.

Quine, W. (1976). Carnap and logical truth, *The Ways of Paradox and Other Essays: Revised and Enlarged Edition*, Harvard University Press, Cambridge, MA, pp. 107–32.

Ramsey, F. (1990). General propositions and causality, in D. Mellor (ed.), *Philosophical Papers*, Cambridge University Press, Cambridge, pp. 145–63.

Restall, G. (1999). *An Introduction to Substructural Logic*, Routledge, London-New York.

Rey, G. (2018). The analytic/synthetic distinction, in E. N. Zalta (ed.), *The Stanford Encyclopedia of Philosophy*, fall 2018 edn, Metaphysics Research Lab, Stanford University, Stanford, CA.

Roberts, C. (2011). Topics, in M. Heusinger and Portner (eds), *Semantics: An International Handbook of Natural Language Meaning*, Vol. 2, De Gruyter Mouton, pp. 1908–34.

Roberts, C. (2012). Information structure: Towards an integrated formal theory of pragmatics, *Semantics and Pragmatics* 5: 1–69.

Rott, H. (2019). Difference-making conditionals and the relevant ramsey test, *Review of Symbolic Logic* Online First: doi 10.1017/S1755020319000674.

Roush, S. (2010). Closure on skepticism, *Journal of Philosophy* 107: 243–56.

Routley, R. (1982). *Relevant Logics and their Rivals: The Basic Philosophical and Semantic Theory*, Ridgeview, Atascadero, CA.

Russell, G. (2008). *Truth in Virtue of Meaning*, Oxford University Press, Oxford.

Ryle, G. (1933). About, *Analysis* 1: 10–12.

Saint-Germier, P. (2020). Hyperintensionality in imagination, in A. Giordani and J. Malinowski (eds), *Logic in High Definition*, Springer, Dordrecht, pp. 77–115.

Salmon, N. (1986). *Frege's Puzzle*, MIT Press, Cambridge, MA.

Schachter, D. and Tulving, E. (1994). *Memory Systems*, MIT Press, Cambridge, MA.

Schipper, A. (2018). Aboutness and negative truths: a modest strategy for truthmaker theorists, *Synthese* 195: 3685722.

Schipper, A. (2020). Aboutness and ontology: A modest approach to truthmakers, *Philosophical Studies* 177: 505–33.

Schipper, B. (2015). Awareness, in H. Van Ditmarsch, J. Halpern, W. van der Hoek and B. Kooi (eds), *Handbook of Epistemic Logic*, College Publications, London, pp. 79–146.

Schroeter, L. (2021). Two-dimensional semantics, in E. N. Zalta (ed.), *The Stanford Encyclopedia of Philosophy*, spring 2021 edn, Metaphysics Research Lab, Stanford University, Stanford, CA.

Scott, D. (1970). Advice on modal logic, in K. Lambert (ed.), *Philosophical Problems in Logic*, Reidel, Dordrecht, pp. 143–73.

Segerberg, K. (1989). Notes on conditional logic, *Studia Logica* 48: 157–68.

Segerberg, K. (1995). Belief revision from the point of view of doxastic logic, *Bulletin of the IGPL* 3: 535–53.

Segerberg, K., Meyer, J.-J. and Kracht, M. (2020). The logic of action, in E. N. Zalta (ed.), *The Stanford Encyclopedia of Philosophy*, summer 2020 edn, Metaphysics Research Lab, Stanford University, Stanford, CA.

Seligman, J. (2014). Situation theory reconsidered, in A. Baltag and S. Smets (eds), *Johan Van Benthem on Logic and Information Dynamics*, Vol. 5 of *Outstanding Contributions to Logic*, Springer, Dordrecht, pp. 895–932.

Sharon, A. and Spectre, L. (2010). Dogmatism repuzzled, *Philosophical Studies* 148: 307–21.

Sharon, A. and Spectre, L. (2017). Evidence and the openness of knowledge, *Philosophical Studies* 174: 1001–37.

Shephard, R. and Metzler, J. (1971). Mental rotation of three-dimensional objects, *Science* 171: 701–3.

Skovgaard-Olsen, N., Singmann, H. and Klauer, K. (2016). The relevance effect and conditionals, *Cognition* 150: 26–36.

Skyrms, B. (2010). *Signals: Evolution, Learning, and Information*, Oxford University Press, Oxford.

Soames, S. (1985). Lost innocence, *Linguistics and Philosophy* 8: 59–71.

Sorensen, R. (1988). Dogmatism, junk knowledge, and conditionals, *Philosophical Quarterly* 38: 433–54.

Sosa, E. (2017). *Epistemology*, Princeton Foundations of Contemporary Philosophy, Princeton University Press, Princeton, NJ.

Speaks, J. (2006). Is mental content prior to linguistic meaning? Stalnaker on intentionality, *Noûs* 40: 428–67.

Spohn, W. (1988). Ordinal conditional functions: A dynamic theory of epistemic states, in L. Hrper and B. Skyrms (eds), *Causation in Decision, Belief Change, and Statistics*, Vol. 2, Kluwer, Dordrecht, pp. 105–34.

Squire, L. (1987). *Memory and Brain*, Oxford University Press, New York.

Stalnaker, R. (1968). A theory of conditionals, in N. Rescher (ed.), *Studies in Logical Theory (American Philosophical Quarterly Monographs 2)*, Blackwell, Oxford, pp. 98–112.

Stalnaker, R. (1975). Indicative conditionals, *Philosophia* 5: 269–86.

Stalnaker, R. (1984). *Inquiry*, MIT Press, Cambridge, MA.

Stalnaker, R. (2003). *Ways a World Might Be: Metaphysical and Anti-Metaphysical Essays*, Oxford University Press, Oxford.

Stanojević, M. (2009). Cognitive synonymy: A general overview, *Facta Universitatis* 7: 193–200.

Starr, W. (2019). Counterfactuals, in E. N. Zalta (ed.), *The Stanford Encyclopedia of Philosophy*, fall 2019 edn, Metaphysics Research Lab, Stanford University, Stanford, CA.

Stenning, K. and van Lambalgen, M. (2008). *Human Reasoning and Cognitive Science*, MIT Press, Cambridge, MA.

Thaler, R. and Sunstein, C. (2008). *Nudge: Improving Decisions about Health, Wealth, and Happiness*, Yale University Press, New Haven, CT.

Troquard, N. and Balbiani, P. (2019). Propositional dynamic logic, in E. N. Zalta (ed.), *The Stanford Encyclopedia of Philosophy*, spring 2019 edn, Metaphysics Research Lab, Stanford University, Stanford, CA.

Van Benthem, J. (2007). Dynamic logic for belief revision, *Journal of Applied Non-Classical Logic* 17: 129–55.

Van Benthem, J. (2011). *Logical Dynamics of Information and Interaction*, Cambridge University Press, Cambridge.

Van Benthem, J. and Martinez, M. (2008). The stories of logic and information, in P. Adriaans and J. Van Benthem (eds), *Handbook of the Philosophy of Information*, Elsevier Science Publishers, Amsterdam, pp. 217–80.

Van Benthem, J. and Smets, S. (2015). Dynamic logics of belief change, in H. Van Ditmarsch, J. Halpern, W. van der Hoek and B. Kooi (eds), *Handbook of Epistemic Logic*, College Publications, London, pp. 313–93.

Van Benthem, J. and Velázquez-Quesada, F. (2010). The dynamics of awareness, *Synthese* 177: 5–27.

Van Ditmarsch, H. (2005). Prolegomena to dynamic logic for belief revision, *Synthese* 147: 229–75.

Van Ditmarsch, H., van der Hoek, W. and Kooi, B. (2008). *Dynamic Epistemic Logic*, Springer, Dordrecht.

Van Fraassen, B. (1969). Facts and tautological entailments, *Journal of Philosophy* 66: 477–87.

Van Fraassen, B. (1980). Review of Brian Ellis, *Rational Belief Systems*, *Canadian Journal of Philosophy* 10: 497–511.

Van Fraassen, B. (1995). Fine-grained opinion, probability, and the logic of full belief, *Journal of Philosophical Logic* 24: 249–77.

Van Leeuwen, N. (2013). The meanings of imagine part i: Constructive imagination, *Philosophy Compass* 8: 220–30.

Van Leeuwen, N. (2016). The imaginative agent, in A. Kind and P. Kung (eds), *Knowledge through Imagination*, Oxford University Press, Oxford, pp. 85–111.

Van Rooij, R. and Schulz, K. (2019). Conditionals, causality and conditional probability, *Journal of Logic Language and Information* 28: 55–71.

Velázquez-Quesada, F. (2011). *Small steps in dynamics of information*, PhD thesis, ILLC, Univerisity of Amsterdam, Amsterdam.

Velázquez-Quesada, F. (2014). Dynamic epistemic logic for implicit and explicit beliefs, *Journal of Logic, Language and Information* 23: 107–40.

Vogel, J. (2014). *e&¬h*, in D. Dodd and E. Zardini (eds), *Scepticism and Perceptual Justification*, Oxford University Press, Oxford, pp. 87–107.

Walton, K. (1990). *Mimesis as Make-Believe*, Harvard University Press, Cambridge, MA.

Wansing, H. (2017). Remarks on the logic of imagination. A step towards understanding doxastic control through imagination, *Synthese* 194: 2843–61.

Weiss, M. and Parikh, R. (2002). Completeness of certain bimodal logics for subset spaces, *Studia Logica* 71: 1–30.

Williamson, T. (2000). *Knowledge and Its Limits*, Oxford University Press, Oxford.

Williamson, T. (2007). *The Philosophy of Philosophy*, Blackwell, Oxford.

Williamson, T. (2016a). Knowing by imagining, in A. Kind and P. Kung (eds), *Knowledge through Imagination*, Oxford University Press, Oxford, pp. 113–23.

Williamson, T. (2016b). Modal science, *Canadian Journal of Philosophy* 46: 453–92.

Wittgenstein, L. (1921/22). *Tractatus logico-philosophicus*, Routledge & Kegan Paul, London.

Yablo, S. (1993). Is conceivability a guide to possibility, *Philosophy and Phenomenological Research* 53: 1–42.

Yablo, S. (2014). *Aboutness*, Princeton University Press, Princeton, NJ.

Yablo, S. (2017). Open knowledge and changing the subject, *Philosophical Studies* 174: 1047–71.

Yablo, S. (2018). Reply to Fine on *Aboutness*, *Philosophical Studies* 175: 1495–512.

Yalcin, S. (2016). Belief as question-sensitive, *Philosophy and Phenomenological Research* 97: 23–47.

Zalta, E. (1988). *Intensional Logic and the Metaphysics of Intentionality*, MIT Press, Cambridge, MA.

Author Index

Subject Index